ELEMENTARY NUMERICAL ANALYSIS

ELEMENTARY NUMERICAL ANALYSIS

by

J. T. Day

VANTAGE PRESS
New York Washington Hollywood

FIRST EDITION

Published by Vantage Press, Inc.
516 West 34th Street, New York, New York 10001

Printed in United States of America

Standard Book No. 533-01046-2

This book is dedicated to
Dr. Preston C. Hammer

PREFACE

This is a book on elementary numerical analysis for undergraduate students in their second and third year. This book is based on a course taught at Penn State University. Except for the chapter on Differential Equations and Integral Equations, all is required is a background of Calculus. It should be possible to cover all of the material in a one semester course. Proofs have been given for most of the results. If the proof is too advanced for this level of course we merely state the results.

It is assumed that a computer is available and the students have a background in Fortran programming. Fortran programs appear here in Fortran IV.

J. T. Day

CONTENTS

ELEMENTARY NUMERICAL ANALYSIS

Chapter I

MATHEMATICAL INTRODUCTION

1.1. A function f(x) is said to be continuous at a point a if whenever $\epsilon > 0$ is given there exists a $\delta > 0$ such that $|x - a| < \delta$ implies $|f(x) - f(a)| < \epsilon$.

A function is continuous on a closed interval [a, b], i.e., $a \leq x \leq b$ if it is continuous at each point of the interval.

Example 1.

Show $f(x) = x^2$ is continuous on [0, 1]. We have for $a \epsilon [0, 1]$, $|x^2 - a^2| = |x - a||x + a| \leq 2|x - a|$; if we chose $\delta = \frac{\epsilon}{2}$ we have $|x^2 - a^2| < \epsilon$.

If a function f(x) is continuous on [a, b] we write $f(x) \epsilon\, c[a, b]$. If f(x), in addition to being continuous, has a continuous first derivative on [a, b] we write $f(x) \epsilon\, c^1[a, b]$. If f(x) and its first n derivatives are continuous on [a, b] we write $f(x) \epsilon\, c^n[a, b]$.

The following theorems are of interest in numerical work.

Theorem 1. If $f(x) \epsilon\, c[a, b]$ then it assumes its maximum and minimum value there.

Theorem 2. If $f(x) \epsilon\, c[a, b]$ and $f(a)\, f(b) < 0$ then there exists a point $\xi\, \epsilon (a, b)$ such that $f(\xi) = 0$.

A theorem that is used repeatedly in numerical analysis is the Mean Value Theorem.

Theorem. Let $f(x) \epsilon\, c^1[a, b]$. Then there exists a number $\xi\, \epsilon (a, b)$ such that

$$f(b) - f(a) = f'(\xi)\, (b - a).$$

Another theorem which has repeated applications is Rolle's Theorem.

Theorem. Let $f(x) \epsilon\, c^1[a, b]$, $f(a) = f(b) = 0$; then there exists a point $\xi \epsilon (a, b)$ such that $f'(\xi) = 0$.

The proof follows from the Mean Value Theorem. We have

$$f(b) - f(a) = (b - a)\, f'(\xi),$$

or

$$0 = f'(\xi)\,(b - a); \text{ hence } f'(\xi) = 0.$$

A theorem of great importance in numerical integration is the following Mean Value Theorem for Integrals.

Theorem. Let $g(x) \epsilon\, c[a, b]$, $f(x) \epsilon\, c[a, b]$. Suppose $g(x)$ doesn't change sign in $[a, b]$, i.e., $g(x) \geq 0$ or $g(x) \leq 0$; then there exists a number $\xi \epsilon (a, b)$ such that

$$\int_a^b g(x)\, f(x) dx = f(\xi) \int_a^b g(x) dx.$$

Example. Consider $\int_{-1}^{1} x^2 \sin x\, dx$. Let $g(x) = x^2$. Since $x^2 \geq 0$ there exists a number ξ such that

$$\int_{-1}^{1} x^2 \sin x\, dx = \sin \xi \int_{-1}^{1} x^2\, dx = \frac{2}{3} \sin \xi$$

where $-1 < \xi < 1$.

A concept of great power and usefulness is that of Taylor's series. Let $f(x) \epsilon\, c^{n+1}[a, b]$. Then

$$f(x) = f(a) + f'(a)\,(x - a) + \frac{f''(a)}{2!}(x - a)^2 + \cdots + \frac{f^n(a)\,(x - a)^n}{n!} + \frac{f^{n+1}(\xi)\,(x - a)^{n+1}}{(n+1)!}$$

where $a < \xi < x$.

Example 2.

Let $f(x) = e^x$, $a = 0$, $b = 1$. Then $f'(x) = e^x$, $f''(x) = e^x$, $f^n(x) = e^x$ for all n. $f(0) = 1$, $f'(0) = 1$, $f''(0) = 1$, $f^n(0) = 1$. The Taylor's series expansion is given by

$$f(x) = e^x = 1 + x + \frac{x^2}{2!} + \frac{x^3}{3!} + \frac{x^4}{4!} + \cdots + \frac{x^n}{n!} + \frac{x^{n+1}}{(n+1)!} e^{\xi}.$$

Example 3.

Use the Taylor expansion through the first three terms to calculate $e^{.05}$, and bound the error made. We have

$$e^x \approx 1 + x + \frac{x^2}{2} \qquad \text{(The "}\approx\text{" means "approximately.")}$$

$$e^{.05} \approx 1 + .05 + \frac{(.05)^2}{2} = 1.05125$$

The error made is $\frac{x^3}{6} e^{\xi}$, $0 < \xi < .05$

$$\frac{(.05)^3}{6} e^{\xi} < \frac{(.05)^3}{6} \cdot e < \frac{(.05)^3}{6} \cdot 3 = .0000625$$

We have used the fact here that $e < 3$.

Example 4.

Let $f(x) = \sin x$. Expand in a Taylor series through the first two non-zero terms about $x = 0$, and bound the error when $x = .05$.

$$f(x) = \sin x,\ f'(x) = \cos x,\ f''(x) = -\sin x$$
$$f'''(x) = -\cos x,\ f^{iv}(x) = \sin x,\ f^{v}(x) = \cos x$$

So

$$f(0) = 0, f'(0) = 1, f''(0) = 0, f'''(0) = -1, f^{iv}(0) = 0, f^{v}(0) = 1$$

$$f(x) = \sin x = x - \frac{x^3}{6} + \frac{x^5}{120} \cos(\xi)$$

$$\sin(.05) \approx .05 - \frac{(0.5)^3}{6} = .0499792$$

The error is bounded by $\left| \frac{(.05)^5}{120} \cos \xi \right| < \frac{(.05)^5}{120}$

Here we used the fact that $|\cos(\xi)| < 1$.

Taylor's series in two variable is as follows:

$$f(x, y) = f(x_0, y_0) + (x - x_0) f_x(x_0, y_0) + (y - y_0) f_y(x_0, y_0)$$
$$+ \frac{1}{2!}\Big[f_{xx}(x_0, y_0)(x - x_0)^2 + 2f_{xy}(x_0, y_0)(x - x_0)(y - y_0)$$
$$+ f_{yy}(x_0, y_0)(y - y_0)^2\Big] + \cdots$$

Example 5.

Expand e^{x+y} in a Taylor's series about $x_0 = 0$, $y_0 = 0$. Through the second order terms,

$$f(x, y) = e^{x+y},\ f_x = e^{x+y},\ f_y = e^{x+y},$$
$$f_{xx} = e^{x+y},\ f_{xy} = e^{x+y},\ f_{yy} = e^{x+y}$$

So

$$f(0, 0) = 1,\ f_x(0, 0) = 1,\ f_y(0, 0) = 1$$
$$f_{xx}(0, 0) = 1,\ f_{xy}(0, 0) = 1,\ f_{yy}(0, 0) = 1.$$

Hence

$$f(x, y) \approx 1 + x + y + \frac{1}{2!}\,[x^2 + 2xy + y^2] + \cdots$$

Exercises

1. Show that $f(x) = x^3$ is continuous on $[0, 1]$.
2. Show $f(x) = x - e^{-x}$ has a zero on the interval $[0, 1]$.
3. Expand $f(x) = \cos(x)$ through fourth order terms about $x_0 = 0$.
4. Calculate $\cos(1)$ using problem 3. Estimate the truncation error.
5. Expand $f(x) = (1 + x)^3$ about $x_0 = 0$ through the first four terms. Compare this with the binomial expansion.
6. Expand $f(x, y) = \cos(x + y)$ about $x_0 = 0$, $y_0 = 0$ through the second order terms.

1.2 Significant Digits. The numbers 0.82525 and 0.0082525 are given to five and seven decimal places respectively. The number of significant digits is defined as the number of digits that are correct starting at the left with the first non-zero digit and proceeding to the right. Thus if all the digits in the two above numbers are correct, we assume they have five significant digits and we write correct to 5D.

If only certain digits in a number are correct, the extra digits may be discarded by rounding off. To do this we round off a number to n significant digits by retaining the n most significant digits and discarding the rest. If a number is less than half a unit in the nth place, we truncate it. If it is greater than or equal to half a unit we increase the nth decimal place by one.

Examples:
3.1415926 rounded to 5D is 3.14159; 8.5350 rounded to 2D is 8.54

Nonlinear Equations

In college algebra one encounters equations of the form $ax + b = 0$. This equation is linear in x because it occurs to the first degree in x. The equation $ax^2 + bx + c = 0$, on the other hand, is nonlinear since it occurs to the second degree in x. In general an equation $f(x) = 0$ is nonlinear if it has terms in x in it which are not of the first degree.

Examples:

$x - e^{-x} = 0$ is nonlinear
$x - \sin x = 0$ is nonlinear
$x - x^3 = 0$ is nonlinear
$x = 3x + 2$ is linear

Chapter 2

NUMERICAL LINEAR ALGEBRA

2.1. In elementary algebra we learned how to solve two equations and two unknowns, for example

(1) $x + y = 3$
(2) $2x - y = 1$

For example, we multiply equation (1) by -2 and add it to equation (2) to obtain $-3y = -5$, $y = \frac{5}{3}$ Then we substitute this value of y in equation (1) to obtain $x + \frac{5}{3} = 3$ to obtain $x = \frac{4}{3}$.

Another way of solving the system is to use Cramer's rule taught in college algebra. We have

$$x = \frac{\begin{vmatrix} 3 & 1 \\ 1 & -1 \end{vmatrix}}{\begin{vmatrix} 1 & 1 \\ 2 & -1 \end{vmatrix}} = \frac{-4}{-3} = \frac{4}{3}$$

$$y = \frac{\begin{vmatrix} 1 & 3 \\ 2 & 1 \end{vmatrix}}{\begin{vmatrix} 1 & +1 \\ 2 & -1 \end{vmatrix}} = \frac{-5}{-3} = \frac{5}{3}.$$

Consider the linear system of equations

$$ax + by = f$$
$$cx + dy = e$$

Then

$$x = \frac{\begin{vmatrix} f & b \\ e & d \end{vmatrix}}{\begin{vmatrix} a & b \\ c & d \end{vmatrix}} = \frac{df - be}{ad - bc}$$

$$y = \frac{\begin{vmatrix} a & f \\ c & e \end{vmatrix}}{\begin{vmatrix} a & b \\ c & d \end{vmatrix}} = \frac{ae - fc}{ad - bc}$$

Here we have assumed that $ad - bc \neq 0$.

By a two-by-two matrix A is meant the square array

$$A = \begin{bmatrix} a & b \\ c & b \end{bmatrix} \quad \text{For example, } A = \begin{bmatrix} 1 & 2 \\ 3 & 4 \end{bmatrix}, A = \begin{bmatrix} 4 & 5 \\ 6 & 7 \end{bmatrix}.$$

The 2 by 1 matrix $x = \begin{bmatrix} x_1 \\ x_2 \end{bmatrix}$ is called a column vector. By multiplication of x by A we mean

$$Ax = \begin{bmatrix} a & b \\ c & d \end{bmatrix} \begin{bmatrix} x_1 \\ x_2 \end{bmatrix} = \begin{bmatrix} ax_1 + bx_2 \\ cx_1 + dx_2 \end{bmatrix}$$

The result is a 2 by 1 vector.

Example 1.

Let $A = \begin{pmatrix} 1 & 2 \\ 3 & 4 \end{pmatrix}$, $x = \begin{pmatrix} 1 \\ 2 \end{pmatrix}$. Then

$$Ax = \begin{pmatrix} 1 & 2 \\ 3 & 4 \end{pmatrix} \begin{pmatrix} 1 \\ 2 \end{pmatrix} = \begin{pmatrix} 1 + 2 \cdot 2 \\ 1 \cdot 3 + 2 \cdot 4 \end{pmatrix} = \begin{pmatrix} 5 \\ 11 \end{pmatrix}.$$

Example 2.

The system of equations

$$ax + by = e$$
$$cx + dy = f$$

can be represented as $AX = B$ where

$$A = \begin{pmatrix} a & b \\ c & d \end{pmatrix} \quad X = \begin{pmatrix} x \\ y \end{pmatrix} \quad B = \begin{pmatrix} e \\ f \end{pmatrix}$$

Thus we have a way of representing linear equations in terms of matrices.

Given two matrices A and B, both two-by-two, we define the matrix product as follows:

$$A \cdot B = \begin{pmatrix} a & b \\ c & d \end{pmatrix} \begin{pmatrix} e & f \\ g & h \end{pmatrix} = \begin{pmatrix} ae + bg & af + bh \\ ce + dg & cf + dh \end{pmatrix}$$

Note that the product of two two-by-two matrices is again a two-by-two matrix.

Example 3.

Let $A = \begin{pmatrix} 1 & 2 \\ 3 & 4 \end{pmatrix}$, $B = \begin{pmatrix} 5 & 6 \\ 7 & 8 \end{pmatrix}$, then

$$A \cdot B = \begin{pmatrix} 1 & 2 \\ 3 & 4 \end{pmatrix} \begin{pmatrix} 5 & 6 \\ 7 & 8 \end{pmatrix} = \begin{pmatrix} 1 \cdot 5 + 7 \cdot 2 & 1 \cdot 6 + 2 \cdot 8 \\ 3 \cdot 5 + 4 \cdot 7 & 3 \cdot 6 + 4 \cdot 8 \end{pmatrix}$$

$$= \begin{pmatrix} 19 & 22 \\ 43 & 50 \end{pmatrix}$$

Example 4.

Compute B · A where B and A are given above

$$\begin{pmatrix} 5 & 6 \\ 7 & 8 \end{pmatrix} \begin{pmatrix} 1 & 2 \\ 3 & 4 \end{pmatrix} = \begin{pmatrix} 5 \cdot 1 + 3 \cdot 6 & 5 \cdot 2 + 6 \cdot 4 \\ 7 \cdot 1 + 3 \cdot 8 & 7 \cdot 2 + 8 \cdot 4 \end{pmatrix} = \begin{pmatrix} 23 & 34 \\ 31 & 46 \end{pmatrix}$$

Thus $BA \neq AB$ for some two-by-two matrices.

It can be shown, however, that $A(BC) = (AB)C$. Thus the associative law holds for two-by-two matrices.

By the identity matrix we mean $I = \begin{pmatrix} 1 & 0 \\ 0 & 1 \end{pmatrix}$. It can be shown that $AI = IA = A$.

By the inverse of the matrix A we mean the two-by-two matrix A^{-1} which, when multiplied by A, yields the identity I, i.e., $A^{-1}A = AA^{-1} = I$.

For a two-by-two matrix the inverse can be explicitly calculated. It is

$$A^{-1} = \frac{\begin{pmatrix} d & -b \\ -c & a \end{pmatrix}}{(ad - bc)}$$

We note that ad − bc $\neq 0$ is a condition for the inverse to exist. This is the determinant by the two-by-two system. Thus the determinant not equal to zero is a condition for the inverse to exist. We define t, a number, times A, a matrix, as $tA = \begin{pmatrix} ta & tb \\ tc & td \end{pmatrix}$.

We have $\begin{pmatrix} 1 & 1 \\ 1 & -1 \end{pmatrix}\begin{pmatrix} x \\ y \end{pmatrix} = \begin{pmatrix} 2 \\ 0 \end{pmatrix}$

Here

$$A = \begin{pmatrix} 1 & 1 \\ 1 & -1 \end{pmatrix},\quad A^{-1} = \frac{\begin{pmatrix} -1 & -1 \\ -1 & 1 \end{pmatrix}}{-2}$$

Hence

$$X = A^{-1}B = \frac{\begin{pmatrix} -1 & -1 \\ -1 & 1 \end{pmatrix}}{-2}\begin{pmatrix} 2 \\ 0 \end{pmatrix} = \frac{\begin{pmatrix} -2 \\ -2 \end{pmatrix}}{-2}$$

or

$$X = \begin{pmatrix} 1 \\ 1 \end{pmatrix}$$

Thus x = 1, y = 1 is the solution.

Exercises

1. Add the matrices A and B where $A = \begin{pmatrix} 2 & 4 \\ 6 & 8 \end{pmatrix}$, $B = \begin{pmatrix} 7 & 10 \\ 12 & 13 \end{pmatrix}$.
2. Find A − B where A and B are given in exercise 1.
3. Find A · B where A and B are given in exercise 1.
4. Find A · B where $A = \begin{pmatrix} 2 & 4 \\ 6 & 8 \end{pmatrix}$, $B = \begin{pmatrix} 7 & 8 \\ 9 & 10 \end{pmatrix}$.
5. Solve the system of equations $2x + y = 3$, $3x - y = 2$ by the matrix method.
6. Solve the system of equations $3x + 2y = 6$, $4x + 7y = 10$ by the matrix method.
7. Solve the system of equations $3x + 6y = 9$, $3x - 6y = -3$ by the matrix method.

8. Write a Fortran program to compute products of two-by-two matrices. In this use the notation $A = \begin{pmatrix} a_{11} & a_{12} \\ a_{21} & a_{22} \end{pmatrix}$, $B = \begin{pmatrix} b_{11} & b_{12} \\ b_{21} & b_{22} \end{pmatrix}$. Then $A \cdot B = C$ where the elements of C, c_{ij} are given by $c_{ij} = \sum_{j=1}^{2} a_{ik} b_{kj}$.

As for a two-by-two matrix it can be shown that $A \cdot B \neq B \cdot A$ for some three-by-three matrices.

By the Identity Matrix I we mean the matrix such that $A \cdot I = I \cdot A = A$.

The identity for a three-by-three matrix is $\begin{bmatrix} 1 & 0 & 0 \\ 0 & 1 & 0 \\ 0 & 0 & 1 \end{bmatrix}$.

Example

$$A = \begin{bmatrix} a_{11} & a_{12} & a_{13} \\ a_{21} & a_{22} & a_{23} \\ a_{31} & a_{32} & a_{33} \end{bmatrix}$$

Then

$$AI = \begin{bmatrix} a_{11} & a_{12} & a_{13} \\ a_{21} & a_{22} & a_{23} \\ a_{31} & a_{32} & a_{33} \end{bmatrix} \begin{bmatrix} 1 & 0 & 0 \\ 0 & 1 & 0 \\ 0 & 0 & 1 \end{bmatrix}$$

$$= \begin{bmatrix} a_{11} \cdot 1 + a_{12} \cdot 0 + a_{13} \cdot 0 & a_{11} \cdot 0 + a_{12} \cdot 1 + a_{13} \cdot 0 & a_{13} \\ a_{21} \cdot 1 + a_{22} \cdot 0 + a_{23} \cdot 0 & a_{21} \cdot 0 + a_{22} \cdot 1 + a_{23} \cdot 0 & a_{23} \\ a_{31} \cdot 1 + a_{32} \cdot 0 + a_{33} \cdot 0 & a_{31} \cdot 0 + a_{32} \cdot 1 + a_{33} \cdot 0 & a_{33} \end{bmatrix}$$

or

$$A \cdot I = \begin{bmatrix} a_{11} & a_{12} & a_{13} \\ a_{21} & a_{22} & a_{23} \\ a_{31} & a_{32} & a_{33} \end{bmatrix} = A$$

To multiply a number times a matrix we multiply each element of the matrix, i.e.,

$$tA = \begin{bmatrix} ta_{11} & ta_{12} & ta_{13} \\ ta_{21} & ta_{22} & ta_{23} \\ ta_{31} & ta_{32} & ta_{33} \end{bmatrix}$$

By the inverse of a three-by-three matrix we mean the matrix A^{-1} such that

$$AA^{-1} = A^{-1}A = I.$$

Example

Solve the system of equations $AX = B$ using the matrix inverse

$$A^{-1}AX = A^{-1}B$$
$$(A^{-1}A)\,X = A^{-1}B$$
$$IX = A^{-1}B$$
$$X = A^{-1}B.$$

So if we could calculate A^{-1} we would have the solution of the three-by-three system of equations.

Consider the matrix $A = \begin{bmatrix} 1 & 2 & 1 \\ 1 & 1 & -1 \\ 1 & -1 & -1 \end{bmatrix}$ to calculate A^{-1},

for this matrix consider the augmented matrix

$$[AI] = \begin{bmatrix} 1 & 2 & 1 & 1 & 0 & 0 \\ 1 & 1 & -1 & 0 & 1 & 0 \\ 1 & -1 & -1 & 0 & 0 & 1 \end{bmatrix}.$$

We now reduce the matrix A in [AI] to the identity matrix by the following operations:

1. Multiply any row by a constant and add it to another row.
2. Interchange rows.
3. Divide any row by a constant.

When we get through we will have [IB]. It can be shown that $B = A^{-1}$.

To calculate A^{-1} we proceed as follows:

Step 1. Multiply row one by -1 and add to rows two and three to obtain

$$\begin{bmatrix} 1 & 2 & 1 & 1 & 0 & 0 \\ 0 & -1 & -2 & -1 & 1 & 0 \\ 0 & -3 & -2 & -1 & 0 & 1 \end{bmatrix}$$

Step 2. Multiply row two by 2 and add to row one, multiply row two by -3 and add to row three.

$$\begin{bmatrix} 1 & 0 & -3 & -1 & 2 & 0 \\ 0 & -1 & -2 & -1 & 1 & 0 \\ 0 & 0 & 4 & 2 & -3 & 1 \end{bmatrix}$$

Step 3. Multiply row two by − 1, and divide row three by 4.

$$\begin{bmatrix} 1 & 0 & -3 & -1 & 2 & 0 \\ 0 & 1 & 2 & 1 & -1 & 0 \\ 0 & 0 & 1 & \frac{1}{2} & -\frac{3}{4} & \frac{1}{4} \end{bmatrix}$$

Step 4. Multiply row three by 3 and add to row one; multiply row three by − 2 and add to row two

$$\begin{bmatrix} 1 & 0 & 0 & \frac{1}{2} & -\frac{1}{4} & \frac{3}{4} \\ 0 & 1 & 0 & 0 & \frac{1}{2} & -\frac{1}{2} \\ 0 & 0 & 1 & \frac{1}{2} & -\frac{3}{4} & \frac{1}{4} \end{bmatrix}$$

$$\begin{bmatrix} 1 & 2 & 1 \\ 1 & 1 & -1 \\ 1 & -1 & -1 \end{bmatrix} \cdot \begin{bmatrix} \frac{1}{2} & -\frac{1}{4} & \frac{3}{4} \\ 0 & \frac{1}{2} & -\frac{1}{2} \\ \frac{1}{2} & -\frac{3}{4} & \frac{1}{4} \end{bmatrix} = \begin{bmatrix} 1 & 0 & 0 \\ 0 & 1 & 0 \\ 0 & 0 & 1 \end{bmatrix}$$

Exercises

1. Solve the system of equations

$$\begin{aligned} x + 4y + z &= 7 \\ x + 4y + 4z &= 1 \\ x + y + z &= 4 \end{aligned}$$

for x, y, z by elimination, by computing the inverse matrix.

2. Given

$$A = \begin{bmatrix} 2 & 3 & 4 \\ 4 & 5 & 6 \\ 7 & 8 & 9 \end{bmatrix} \quad B = \begin{bmatrix} 1 & 1 & 1 \\ 1 & 2 & 1 \\ 1 & 3 & 1 \end{bmatrix}$$

Compute $A \cdot B$, $A + B$, $A + 2B$.

3. Compute B^{-1} where B is given by

$$B = \begin{bmatrix} 1 & 1 & 1 \\ 1 & 2 & 1 \\ 0 & 2 & 1 \end{bmatrix}$$

2.2 Gaussian elimination. A method for solving systems of equations which is faster on a computer than finding the inverse matrix is a method due to Gauss. We explain it for a three-by-three system of equations.

Let the system

$$a_{11}x_1 + a_{12}x_2 + a_{13}x_3 = a_{14}$$
$$a_{21}x_1 + a_{22}x_2 + a_{23}x_3 = a_{24}$$
$$a_{31}x_1 + a_{32}x_2 + a_{33}x_3 = a_{34} \qquad (1)$$

be given with its determinant non-zero. Without loss of generality, we assume $a_{11} \neq 0$. (If $a_{11} = 0$, reorder equations for some a_{i1} is not zero). Divide coefficients of the first equation by a_{11} to obtain a new equation

$$x_1 + b_{12}x_2 + b_{13}x_3 = b_{14}. \qquad (2)$$

Here $b_{1k} = \dfrac{a_{1k}}{a_{11}}$ $(k = 2, 3, 4)$.

We now add equation (2), multiplied by $-a_{21}$ to the second equation of system (1), and equation (2) multiplied by $-a_{31}$, to the third equation of system (1). We obtain a system of two equations in two unknowns, x_2 and x_3

$$a_{22}^{(1)} x_2 + a_{23}^{(1)} x_3 = a_{24}^{(1)}$$

$$a_{32}^{(1)} x_2 + a_{33}^{(1)} x_3 = a_{34}^{(1)} \qquad (3)$$

Here the superscript signifies that we are talking about the transformed equation.

The a_{ik}^{1} are computed by

$$a_{ik}^{1} = a_{ik} - a_{i1}\, b_{1k} \qquad (i = 2, 3, \text{ and } k = 2, 3, 4).$$

If $a_{22}^{1} \neq 0$ (either $a_{22}^{(1)} \neq 0$ or $a_{32}^{(1)} \neq 0$), then we divide through by a_{22}^{1} to get the equation

$$x_2 + b_{23}^{(1)} x_3 = b_{24}^{(1)} \qquad (4)$$

where $b_{2k}^{1} = \dfrac{a_{2k}^{(1)}}{a_{22}^{(1)}}$ $(k = 3, 4)$

We can now eliminate the coefficient of the unknown x_2 from system (3) by using system (4) and reduce it to a single equation

$$a_{33}^{(2)} x_3 = a_{34}^{(2)}$$

where $a_{3k}^{(2)} = a_{3k}^{(1)} - a_{32}^{(1)} b_{2k}^{(1)}$. Assuming $a_{33}^{2} \neq 0$, we get

$$x_3 = \frac{a_{34}^{(2)}}{a_{33}^{(2)}} = b_{34}^{(2)} \tag{5}$$

Combining equations (2), (4) and (5), we obtain the system

$$\begin{aligned} x_1 + b_{12}x_2 + b_{13}x_3 &= b_{14} \\ x_2 + b_{23}^{(1)} x_3 &= b_{24}^{(1)} \\ x_3 &= b_{34}^{(2)} \end{aligned} \tag{6}$$

Thus we have triangulated the system. We solve for x_1 and x_2 and x_3 as follows:

$$\begin{aligned} x_3 &= b_{34}^{(2)} \\ x_2 &= b_{24}^{(1)} - b_{23}^{(1)} x_3 \\ x_1 &= b_{14} - b_{13} x_3 - b_{12} x_2 \end{aligned} \tag{7}$$

Finding the solution using (7) is called the process of backward substitution.

An example of the above is as follows

$$\begin{aligned} 2x_1 + 3x_2 + 5x_3 &= 10 \\ 3x_1 + 4x_2 + 6x_3 &= 13 \\ 5x_1 + 10x_2 + 5x_3 &= 20 \end{aligned} \tag{8}$$

We divide by $a_{11} = 2$ to obtain

$$x_1 + \frac{3}{2} x_2 + \frac{5}{2} x_3 = 5 \tag{9}$$

Then eliminate a_{21} and a_{31} by multiplying (9) by -3 and -5 respectively and adding to obtain

$$x_1 + \frac{3}{2}x_2 + \frac{5}{2}x_3 = 5$$

$$-\frac{1}{2}x_2 - \frac{3}{2}x_3 = -2$$

$$\frac{5}{2}x_2 - \frac{15}{2}x_3 = -5 \qquad (10)$$

We then divide the second equation by $-\frac{1}{2}$ to obtain

$$x_2 + 3x_3 = 4 \qquad (11)$$

We eliminate out the x_2 term in the third equation by multiplying (11) by $-\frac{5}{2}$ and adding to obtain

$$-15x_3 = -15$$

or

$$x_3 = 1$$

Then

$$x_2 = 4 - 3x_3 = 4 - 3 = 1$$

$$x_3 = -\frac{3}{2}x_2 - \frac{5}{2}x_3 + 5 = 1$$

A Fortran program for Gaussian elimination follows.

```
      DIMENSION A(3, 4), B(3, 4), X(3), AP(3, 4)
      DIMENSION AP2(3, 4)
      READ (5, 1) ((A(I, J), J = 1, 4), I = 1, 3)
1     FORMAT (4F10.3)
            DO 2K = 2, 4
2     B(1, K) = A(1, K)/A(1, 1)
            DO 3I = 2, 3
            DO 3K = 2, 4
3     AP (I, K) = A(I, K) - A(I, 1) * B(I, K)
            DO 4K = 3, 4
4     B(2, K) = AP(2, K)/AP(2, 2)
            DO 5K = 3, 4
```

```
 5    AP2(3, K) = AP(3, K) - AP(3, 2) * B(2, K)
      B(3, 4) = AP2(3, 4)/AP(3, 3)
      X(3) = B(3, 4)
      X(2) = B(3, 4) - B(2, 3) * X(3)
      X(1) = B(1, 4) - B(1, 3) * X(3) - B(1, 2) * X(2)
      WRITE (6, 10) (X(I), I = 1, 3)
10    FORMAT (3E20.8)
      STOP
      END
```

Exercises

1. Solve exercise 1, page 22, by Gaussian elimination.
2. Solve the system

$$3x_1 + 10x_2 + 5x_3 = 21$$
$$10x_1 - 10x_2 + 5x_3 = 15$$
$$30x_1 + 10x_2 + 5x_3 = 75$$

   by Gaussian elimination.
3. Generalize the algorithm of this section to 4 equations and 4 unknowns and write a Fortran program to solve the system.

Chapter 3

THE SOLUTION OF NONLINEAR EQUATIONS BY ITERATION

3.1 Introduction. In this section we shall consider the solution of nonlinear equations of the form $f(x) = 0$ by iterative methods.

We write the equation $f(x) = 0$ in the form $x = g(x)$. For example, we write $f(x) = x^2 - 3x + 2 = 0$ in the form $x = \frac{x^2 + 2}{3}$, for the purpose of obtaining the following algorithm.

Algorithm 3.1 Let x_0 be given; then generate the sequence x_n according to $x_n = g(x_{n-1})$.

For example, let $g(x) = \frac{x^2 + 2}{3}$, $x_0 = 0$.

Then

$$x_1 = \frac{0 + 2}{3} = \frac{2}{3} = .66666667$$

$$x_2 = \frac{\left(\frac{2}{3}\right)^2 + 2}{3} = \frac{22}{27} = .81481481$$

$$x_3 = \frac{\left(\frac{22}{27}\right)^2 + 2}{3} = \frac{1942}{2187} = .88797439$$

$$x_4 = \frac{x_3^2 + 2}{3} = .92949951$$

We ask the question: does the limit of x_n as $n \to \infty$ exist and if so, what is the relationship between the limit and the solution of the equation $x - g(x) = 0$. In the succeeding section we shall prove some theorems which answer the above question.

Section 3.2, Theorem 3.1. Let $g(x) \in c[a, b]$ and let its range be included in its domain. Then the equation $x = g(x)$ has a solution.

If $a = g(a)$ or $b = g(b)$ then we are through. Suppose this is not the case. Form $F(t) = t - g(t)$. Then $F(a) = a - g(a)$ and $F(a) < 0$, $F(b) = b - g(b)$ and $F(b) > 0$. Since $F(t)$ is continuous we may use the theorem "If $F(t) \in c[a, b]$ and $F(a)\,F(b) < 0$ then there exists a zero s in $[a, b]$." Therefore we have a value s such that $F(s) = 0$, or $s = g(s)$.

Example. Let $[a, b] = [0, 1.1]$ then if $g(x) = \dfrac{x^2 + 2}{3}$, $g(x) \in \left[\dfrac{2}{3}, 1.07\right]$ c[a, b] so there exists a zero s of $s = g(s)$ in $[a, b]$.

In numerical work we need not only know that a zero exists on a certain interval but also that the zero is unique, if the iterative process $x_n = g(x_{n-1})$ is to work. The following theorem guarantees that we have a unique solution to the equation $x = g(x)$ in a certain interval.

Theorem 3.2. Let $g(x) \in c^1\,[a, b]$ and let its range be included in its domain. Moreover suppose there exists a number $r < 1$ such that $|g'(x)| < r$ for $x \in [a, b]$ then the equation $x = g(x)$ has a unique solution in $[a, b]$.

By Theorem 3.1 there exists a zero s of $x = g(x)$ in $[a, b]$. Suppose there were two, say s and t. Then $s = g(s)$, $t = g(t)$. Subtracting the second from the first we have $s - t = g(s) - g(t) = (s - t)\,g'(\xi)$, where $\xi \in [a, b]$. Since $s \neq t$, we can divide by $s - t$ to obtain $1 = g'(\xi)$. However this is a contradiction, since $|g'(x)| < r < 1$. Thus we have shown that the zero of $x = g(x)$ is unique.

Example.

$$x = g(x) = \frac{x^2 + 2}{3},\ x \in [0, 1.1] = [a, b].$$

$g(x) \in [a, b]$ was shown above.

$$g'(x) = \frac{2x}{3},\ |g'(x)| \leq \frac{2 \cdot 1.1}{3} < .734$$

Therefore, $r = .734$ and the example has a unique solution in $[0, 1.1]$.

The next theorem insures us under what conditions the iterative method $x_n = g(x_{n-1})$ converges to s.

Theorem 3.3. Let $g(x) \epsilon c^1[a, b]$ and let its range be included in its domain; moreover let there exist a number $r < 1$ such that $|g'(x)| < r < 1$ for $x \epsilon [a, b]$. Then for $x_0 \epsilon [a, b]$ the iteration $x_n = g(x_{n-1})$ converges to the unique solution s of $x = g(x)$, i. e., $s = g(s)$.

First we must show that $x_n \epsilon [a, b]$, $x_1 = g(x_0)$ and $g(x_0) \epsilon [a, b]$. Therefore $x_1 \epsilon [a, b]$. By induction assume $x_n \epsilon [a, b]$. Then $x_{n+1} = g(x_n)$, and $g(x_n) \epsilon [a, b]$; therefore $x_{n+1} \epsilon [a, b]$. By the principle of finite induction $x_n \epsilon [a, b]$ for all n.

$$x_1 = g(x_0)$$
$$s = g(s)$$
$$x_1 - s = g(x_0) - g(s) = (x_0 - s)\, g'(\xi_1)$$
$$|x_1 - s| < |x_0 - s|\, r$$
$$x_2 = g(x_1)$$
$$s = g(s)$$
$$x_2 - s = g(x_1) - g(s) = (x_1 - s)\, g'(\xi_2)$$
$$|x_2 - s| < |x_1 - s|\, r < r^2 |x_0 - s|$$

Continuing inductively, we find that

$$|x_n - s| < r^n |x_0 - s|$$

Taking the limit at $n \rightarrow \infty$ we find $\lim r^n \rightarrow 0$ since $r < 1$. Therefore

$$\lim |x_n - s| \rightarrow 0 \text{ or } \lim_{n \rightarrow \infty} x_n = s.$$

In our example $x = g(x) = \dfrac{x^2 + 2}{3}$ we find for $x \epsilon [0, 1.1]$. We know g(x) satisfies the hypothesis of the theorem, so $x_n \rightarrow 1$.

Our second example is $F(x) = x - \cos(x) = 0$, written in the form $x = \cos(x) = g(x)$. Take $[a, b] = [.7, .79]$, then

$$g(x) \epsilon [.70385, .76484] \epsilon [a, b]$$
$$g'(x) = -\sin x \text{ so } |g'(x)| \leq .71035 \text{ or } r = .72$$

So we have a unique solution. Taking $x_0 \epsilon .74$ we iterate ten times to obtain

$$x_1 = .73846855$$
$$x_2 = .73950032$$
$$x_3 = .73880539$$
$$x_4 = .73927354$$
$$\vdots$$
$$x_{10} = .73910273$$

Our last example is $x = e^{-\frac{x}{2}} = g(x)$. Take $[a, b] = [0, 1]$. We have $g(x) \epsilon [e^{-\frac{1}{2}}, 1] = [.606531, 1] \subset [a, b]$. $g'(x) = -\frac{1}{2} e^{-\frac{x}{2}}$ so $| g'(x) | \leqq \frac{1}{2}$.

Hence the iteration will converge if $x_0 \epsilon [0, 1]$. Choosing $x_0 = .5$ we obtain the following five interates

$$x_1 = .77880078$$
$$x_2 = .67746297$$
$$x_3 = .71267379$$
$$x_4 = .70023667$$
$$x_5 = .70460470$$

A Fortran Program to iterate $x = e^{-\frac{x}{2}}$, $x_0 = .5$, $N = 10$ is as follows:

```
      F(X) = EXP(- X/2.)
      X = .5
      N = 10
      IT = 0
      WRITE (6, 10) IT, X
      DO 1 I = 1, N
      X = F(X)
      WRITE (6, 10, I, X)
   10 FORMAT (I10, E20.8)
    1 CONTINUE
      STOP
      END
```

A geometric picture of how the iterative method works is as follows:

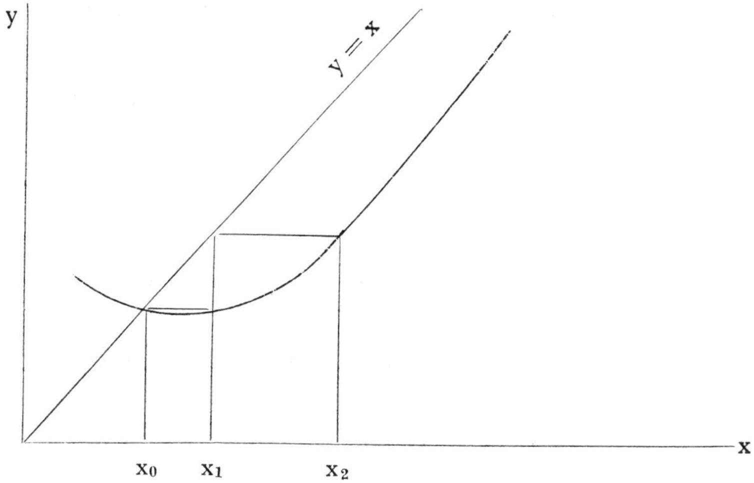

Exercises

1. Find an interval in which the iteration $x_{n+1} = g(x_n)$ converges where $x = g(x) = \dfrac{x^2 + 5}{6}$.

2. Show the iteration $x = g(x) = .9 + .1 \sin x$ will converge for any $x_0 \in [-\pi, \pi]$.

3. Find an interval about $x = 2$ so that the iteration $x = g(x) = \dfrac{1}{2}\left(x + \dfrac{4}{x}\right)$ will converge.

4. Write a Fortran program to iterate problem 1 ten times.

5. Write a Fortran program to iterate problem 2 ten times.

Section 3. 3. Newton's Method. We noticed in our examples in section 2.2 that convergence may be quite slow for the method considered there. In this section we shall develop a method which in some circumstances converges more rapidly to the solution.

Let s be a zero of $f(x) = 0$. Let x_n be a guess in an iterative procedure to be developed. Expanding $f(s)$ in Taylor's series about x_n we obtain $f(s) = f(x_n) + (s - x_n) f'(x_n) +$ higher order terms (3.2)

Now $f(s) = 0$, and neglecting the higher order terms, we solve equation (3.2) for s to obtain

$$s = x_n - \frac{f(x_n)}{f'(x_n)}.$$

Since we neglected the higher order terms we will get just an approximation to x. Call it x_{n+1}. We have the following algorithm :

Algorithm 3.2. Let x_0 be given. Generate the x_n according to the scheme

$$x_{n+1} = x_n - \frac{f(x_n)}{f'(x_n)}$$

Consider the problem of finding the $\sqrt{N}$. Then $f(x) = x^2 - N$ so Newton's Method yields

$$f'(x) = 2x$$

$$x_{n+1} = x_n - \frac{x_n^2 - N}{2x_n} = \frac{1}{2}\left(x_n + \frac{N}{x_n}\right)$$

For $N = 9$ we have $x_{n+1} = \frac{1}{2}\left(x_n + \frac{9}{x_n} \right)$. Let $x_0 = 2$; we have then

$$x_1 = 3.25$$
$$x_2 = 3.009618845$$
$$x_3 = 3.000015371$$
$$x_4 = 3.000000000$$

Thus after four iterations the method has converged to 9 significant digits.

Suppose we wish to find the solution of the equation $f(x) = \frac{1}{x} - A = 0.$

The solution of this equation is $\frac{1}{A}$.

$$f'(x) = -\frac{1}{x^2}$$

$$x_{n+1} = x_n - \frac{\frac{1}{x_n} - A}{-\frac{1}{x_n^2}} = x_n + x_n^2\left(\frac{1}{x_n} - A\right)$$

$$= 2x_n^2 - Ax_n^2 = x_n(2 - Ax_n)$$

Let $A = 7$ so we are calculating $\frac{1}{7}$. Let $x_0 = .1$. We find

$x_1 = .13$
$x_2 = .1417$
$x_3 = .14284777$
$x_4 = .142857142242$

whereas the exact value is .142857142857 to 12 D.

We shall now investigate Newton's method for the square root. In particular we shall say that for any $x_0 > 0$ that $x_n \to \sqrt{N}$ as $n \to \infty$.

Newton's method for the square root is $x_{n+1} = \frac{1}{2}\left(x_n + \frac{N}{x_n}\right)$. Thus we have $x_1 = \frac{1}{2}\left(x_0 + \frac{N}{x_0}\right)$. Subtracting $\sqrt{N}$ from both sides of the equation we have

$$x_1 - \sqrt{N} = \frac{1}{2}\left(x_0 + \frac{N}{x_0}\right) - \sqrt{N} = \frac{x_0^2 - 2\sqrt{N}\,x_0 + N}{2x_0}$$

or

$$x_1 - \sqrt{N} = (x_0 - \sqrt{N})^2/2x_0$$

Similarly $x_1 + \sqrt{N} = (x_0 + \sqrt{N})^2/2x_0$.

Thus

$$\frac{x_1 - \sqrt{N}}{x_1 + \sqrt{N}} = \left(\frac{x_0 - \sqrt{N}}{x_0 + \sqrt{N}}\right)^2.$$

We next establish by induction that

$$\frac{x_n - \sqrt{N}}{x_n + \sqrt{N}} = \left(\frac{x_0 - \sqrt{N}}{x_0 + \sqrt{N}}\right)^{2^n}$$

The case n = 1 was established above. Now

$$\frac{x_{n+1} - \sqrt{N}}{x_{n+1} + \sqrt{N}} = \left(\frac{x_n - \sqrt{N}}{x_n + \sqrt{N}}\right)^2$$

as can be seen in a similar manner as the x_0, x_1 case above. So

$$\frac{x_{n+1} - \sqrt{N}}{x_{n+1} + \sqrt{N}} = \left(\frac{x_0 - \sqrt{N}}{x_0 + \sqrt{N}}\right)^{2^n} \left(\frac{x_0 - \sqrt{N}}{x_0 + \sqrt{N}}\right)^{2^n}$$

$$= \left(\frac{x_0 - \sqrt{N}}{x_0 + \sqrt{N}}\right)^{2^n + 2^n} = \left(\frac{x_0 - \sqrt{N}}{x_0 + \sqrt{N}}\right)^{2^n \cdot 2}$$

$$= \left(\frac{x_0 - \sqrt{N}}{x_0 + \sqrt{N}}\right)^{2^{n+1}}$$

Thus the relationship is established for all n.

Now $|x_0 - \sqrt{N}| / |x_0 + \sqrt{N}| < 1$ since $x_0 > 0$.

Let $|x_0 - \sqrt{N}| / |x_0 + \sqrt{N}| = r$.

Then we have

$$\frac{x_n - \sqrt{N}}{x_n + \sqrt{N}} = r^{2^n}$$

Since $\lim_{n \to \infty} r^{2^n} \to 0$ we have $x_n \to \sqrt{N}$.

We next state a theorem on the global convergence of Newton's method. For a proof of the theorem see Henrici [ ].

Theorem 3.4. Let $f(x) \in c^2 [a, b]$ and let the following conditions be satisfied :

1. $f(a) f(b) < 0$
2. $f'(x) \neq 0 \; x \in [a, b]$
3. $f''(x) \geq 0$ or $f''(x) \leq 0$ for $x \in [a, b]$
4. If t denotes the endpoint of [a, b] at which $|f'(x)|$ is smaller then

$$\left|\frac{f(t)}{f'(t)}\right| \leq b - a .$$

Then Newton's method converges to the solution s of $f(s) = 0$ for any $x_0 \in [a, b]$.

Condition one states that there is a solution s of $f(s) = 0$ in $[a, b]$. Condition two states there is only one solution, since if for two distinct p, q $f(p) - f(q) = (p - q)\, f'(\xi)$ or $0 = f'(\xi)$, a contradiction. Condition three states that this graph of $f(x)$ is either concave from above or below.

Example. Apply the theorem to the equation $f(x) = x^2 - 3x + 2 = 0$. We put $[a, b] = [0, 1.4]$. Then $f(0) = 2$, $f(1.4) = -.24$, so condition one is satisfied. Now $f'(x) = 2x - 3 \neq 0$ for $x \in [0, 1.4]$ and $f'''(x) = 2$ so condition three is satisfied.

$$| f'(0) | = 3, \; | f'(1.4) | = .2 \text{ so } t = 1.4$$

and

$$\left| \frac{f(1.4)}{f'(1.4)} \right| = \frac{.24}{.2} = 1.2 .$$

Since $b - a = 1.4$ we have

$$\left| \frac{f(1.4)}{f'(1.4)} \right| = 1.2 < 1.4 .$$

Hence condition four is satisfied. Thus, Newton's method will converge for any $x_0 \in [0, 1.4]$ to the root $x = 1$.

Write a Fortran program to find the zero between 2 and 3 of $f(x) = x^3 - 2x - 5$, using Newton's method. Here $f' = 3x^2 - 2$, $x_0 = 2$, $N = 25$.

```
      F(X) = X**3 - 2. *X - 5.
      DF(X) = 3.*X*X - 2.
      X = 2.
      IT = 0
      WRITE (6, 10) IT, X
      N = 25
      DO 1 I = 1, N
      X = X - F(X)/DF(X)
      WRITE (6, 10) I, X
   10 FORMAT (I10, E20 .8)
    1 CONTINUE
      STOP
      END
```

Geometrically Newton's method can be explained as follows: We take the tangent line at x_0 to f(x), let it intersect the x-axis, call this x_1, etc.
A graph is as follows:

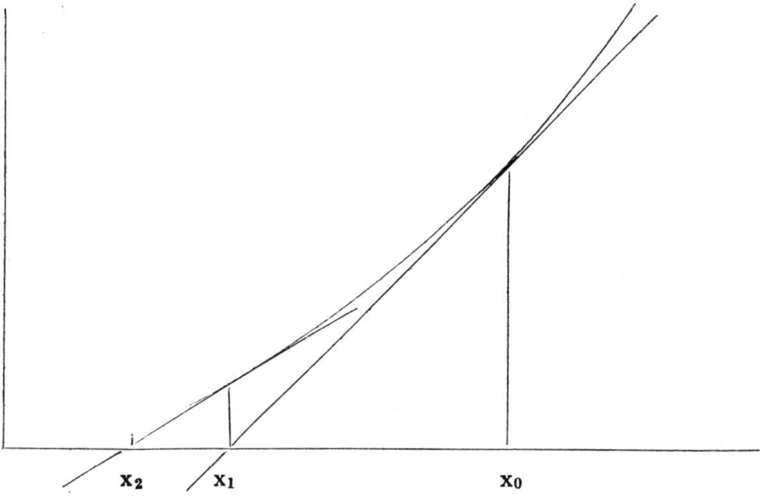

Exercises

1. Find an interval in which Newton's method will converge for $f(x) = x^2 - 2 = 0$, using Theorem 3.2.
2. Calculate $\frac{1}{10}$ using the method discussed in this section; use $x_0 = .05$.
3. Calculate the zero of $f(x) = x - \cos x$ to 4 decimal places using $x_0 = .7$ by Newton's method.
4. Let $f(x) = a + bx$. How many iterations are necessary for Newton's method to converge to the zero ? Let $x_0 = 0$.
5. Calculate $\sqrt{5}$ to 5 decimal places using Newton's method, with $x_0 = 2$.
6. Find an interval in which Newton's method converges to the smaller root of $f(x) = x^2 - 5x + 6 = 0$.
7. Calculate $\sqrt{e}$ to 5 decimal places with $x_0 = 1.5$.
8. Determine to two decimal places the root between 2 and 3 of $f(x) = x^3 - 2x - 5 = 0$ by using Newton's method.

Section 3.4 Newton's Method for Functions of Two Variables. In this section we discuss the iterative solution of the system of equations $f(x, y) = 0$, $g(x, y) = 0$ by the generalization of Newton's method to functions of two variables.

Suppose the point (s, t) is a solution of the above system, i.e., $g(s, t) = f(s, t) = 0$. Then take some point (x_n, y_n) near (s, t) and expand f and g in Taylor's series about the point (x_n, y_n). We obtain

$$f(s, t) = f(x_n, y_n) + f_x(x_n, y_n)(s - x_n) + f_y(x_n, y_n)(t - y_n) + \text{higher order terms} \qquad (1)$$

$$g(s, t) = g(x_n, y_n) + g_x(x_n, y_n)(s - x_n) + g_y(x_n, y_n)(t - y_n) + \text{higher order terms} \qquad (2)$$

If we neglect the higher order terms in equations (1) and (2) we see that we have a system of linear equations to be solved for (s, t).

$$0 = f(x_n, y_n) + f_x(x_n, y_n)(s - x_n) + f_y(x_n, y_n)(t - y_n) \qquad (3)$$

$$0 = g(x_n, y_n) + g_x(x_n, y_n)(s - x_n) + g_y(x_n, y_n)(t - y_n) \qquad (4)$$

Since we neglected the higher order terms s and t are not found exactly; instead (x_{n+1}, y_{n+1}) are found. We thus have the system of equations (5) and (6).

$$0 = f(x_n, y_n) + f_x(x_n, y_n)(x_{n+1} - x_n) + f_y(x_n, y_n)(y_{n+1} - y_n) \qquad (5)$$

$$0 = g(x_n, y_n) + g_x(x_n, y_n)(x_{n+1} - x_n) + g_y(x_n, y_n)(y_{n+1} - y_n) \qquad (6)$$

Thus we have a matrix equation, letting $P_n = (x_n, y_n)$

$$\begin{bmatrix} f_x(P_n) & f_y(P_n) \\ g_x(P_n) & g_y(P_n) \end{bmatrix} \begin{bmatrix} x_{n+1} - x_n \\ y_{n+1} - y_n \end{bmatrix} = - \begin{bmatrix} f(P_n) \\ g(P_n) \end{bmatrix}$$

to be solved for (x_{n+1}, y_{n+1}). Inverting the two-by-two matrix we have

$$\begin{bmatrix} x_{n+1} - x_n \\ y_{n+1} - y_n \end{bmatrix} = \frac{\begin{bmatrix} g_y(P_n) & -f_y(P_n) \\ -g_x(P_n) & f_x(P_n) \end{bmatrix}}{(f_x(P_n)\, g_y(P_n) - g_x(P_n)\, f_y(P_n))} \begin{bmatrix} f(P_n) \\ g(P_n) \end{bmatrix}$$

or

$$x_{n+1} = x_n - \frac{(g_y(P_n)\, f(P_n) - f_y(P_n)\, g(P_n))}{(f_x(P_n)\, g_y(P_n) - g_x(P_n)\, f_y(P_n))} \qquad (7)$$

$$y_{n+1} = y_n - \frac{(f_x(P_n)\, g(P_n) - g_x(P_n)\, f(P_n))}{(f_x(P_n)\, g_y(P_n) - g_x(P_n)\, f_y(P_n))} \qquad (8)$$

Exercises

1. Solve the system $f(x,y) = x^2 + y^2 - 25 = 0$; $g(x,y) = y - x = 0$, for the roots of the equation near (3.5, 3.5).
2. Solve the system $f(x, y) = x^2 + 4y^2 - 36 = 0$; $g(x, y) = x^2 + y^2 - 12x + 27 = 0$ with $x_0 = 0$, $y_0 = -8$.

Section 3.5 The Method of Bisection. In this section we shall consider the method of bisection of determining the root of an equation. We illustrate the method.

Consider the problem of finding the root of the equation $x^3 - 2x - 5 = 0$ between one and three.

We have $P(1) = 1 - 2 - 5 = -6$, $P(3) = 27 - 6 - 5 = 16$, so $P(x)$ has a root between 1 and 3.

We halve the interval, $s = \frac{1 + 3}{2} = 2$

$P(2) = 8 - 4 - 5 = -1$, so a root is between 2 and 3.
$s = \frac{2 + 3}{2} = 2.5$

$P(2.5) = 15.625 - 5 - 5 = 6.25$. So a root lies between 2 and 2.5. $s = \frac{2 + 2.5}{2} \cong 2.3.$

$P(2.3) = 12.167 - 4.6 - 5 > 0$, so a root lies between 2 and 2.3. $s = \frac{2 + 2.3}{2} \cong 2.2.$

$P(2.2) = 10.758 - 4.4 - 5 > 0$, so a root lies between 2 and 2.2. $s = \frac{2 + 2.2}{2} = 2.1$

$P(2.1) = 9.261 - 4.2 - 5 = .061 > 0$. Hence a root lies between 2 and 2.1. $s = \frac{2 + 2.1}{2} = 2.05.$

$P(2.05) = 8.615125 - 4.10 - 5 < 0$, so a root lies between 2.05 and 2.1.

Then Newton's method may be used to find the root.

Exercises

1. Write a Fortran program using the method of bisection to compute the real positive root of $x^3 - 2x - 5 = 0$. Use Newton's method when you have the root to within .005. Calculate the root to six decimal places.
2. Use the method of bisection on $x^2 - 2x + 1 = 0$ with initial estimates 0 and 3. Compute the root correct to 10 decimal places.
3. Determine to 2D the two real roots of $3x^4 + 8x^3 - 6x^2 - 25x + 19 = 0$ by the method of bisection.
4. Use 0 and 2 as initial estimates of the root to the equation $x^3 - x^2 - x + 1 = 0$. Find all three roots by the method of bisection.

Section 3.6 The Method of Regulai Falsi. In Newton's method we have $x_{k+1} = x_k - \dfrac{f(x_k)}{f'(x_k)}$. It may be that $f'(x)$ is hard to evaluate or doesn't exist.

In either case it may be opportune to replace $f'(x_k)$ by $\dfrac{f(x_k) - f(x_{k-1})}{x_k - x_{k-1}}$ to obtain $x_{k+1} = x_k - \dfrac{(x_k - x_{k-1})}{(f(x_k) - f(x_{k-1}))} f(x_k)$. This is the method of Regulai Falsi.

Example. Apply the above material on Regulai Falsi to $f(x) = x - e^{-x} = 0$, with $x_0 = .5$, $x_1 = .6$. Now

$$x_2 = x_1 - \frac{(x_1 - x_0)}{f(x_1) - f(x_0)} f(x_1)$$

$$= .6 - \frac{(.1)\,(.6 - .548812)}{(.6 - .548812 - (.5 - .606531))}$$

$$= .6 - \frac{(.1)\,(.051188)}{.157719} = .6 - .032 = .568$$

Example. A Fortran program to compute the root of $x - e^{-x} = 0$ with the method of Regulai Falsi follows:

```
F(X) = X - EXP(- X)
XZ = .5
X1 = .6
IT = 0
```

```
      WRITE (6, 10) IT, X
      WRITE (6, 10) IT, X1
      N = 10
      DO 1I = 1, N
      X2 = X1 - (X1 - XZ) * F(X1)/F(X1) - F(XZ))
      WRITE (6, 10) I, X2
   10 FORMAT (I10, E20.8)
      XZ = X1
      X1 = X2
      IF(ABS(X1 - XZ). LE .0000001) STOP
    1 CONTINUE
      STOP
      END
```

Exercises

1. Use the method of Regulai Falsi to compute to 2D the root of x − cos(x) = 0.
2. Write a Fortran program to compute to 5D using the method of Regulai Falsi for the equation x − cos(x) = 0.

Chapter 4

INTERPOLATION

Section 4.1 Linear Interpolation. In this section we shall study linear interpolation, a process that is studied in an elementary course on trigonometry and logarithms.

Given a continuous function on $[x_0, x_1]$ we fit a straight line $P(x)$ to the data so that $P(x_0) = f(x_0)$, and $P(x_1) = f(x_1)$. If the straight line is represented as $P(x) = Ax + B$, we must solve the system of equations

$$P(x_0) = f(x_0) = Ax_0 + B$$
$$P(x_1) = f(x_1) = Ax_1 + B$$

for A and B. This can be done provided the determinant of coefficients is nonzero. In this case

$$\begin{vmatrix} x_0 & 1 \\ x_1 & 1 \end{vmatrix} = x_0 - x_1 \neq 0 \text{ if } x_1 \neq x_0.$$

So one can always solve for A and B provided that the points x_0 and x_1 are distinct.

$$A = \frac{\begin{vmatrix} f(x_0) & 1 \\ f(x_1) & 1 \end{vmatrix}}{(x_0 - x_1)} = \frac{f_0 - f_1}{x_0 - x_1}$$

$$B = \frac{\begin{vmatrix} x_0 & f_0 \\ x_1 & f_1 \end{vmatrix}}{(x_0 - x_1)} = \frac{x_0 f_1 - x_1 f_0}{(x_0 - x_1)}$$

Thus

$$P(x) = \frac{f_0 - f_1}{x_0 - x_1} x + \frac{x_0 f_1 - x_1 f_0}{x_0 - x_1}$$

$$P(x) = \frac{(x - x_1)}{(x_0 - x_1)} f_0 + \frac{(x - x_0)}{(x_1 - x_0)} f_1$$

Example 1.

Show

$$\frac{x - x_1}{x_0 - x_1} + \frac{x - x_0}{x_1 - x_0} = 1$$

We have

$$\frac{x_1 - x}{x_1 - x_0} + \frac{x - x_0}{x_1 - x_0} = \frac{x_1 - x_0}{x_1 - x_0} = 1$$

This gives us a simple computational check on our results.

Example 2.

Given $f(x) = \sin(x)$, $f(.1) = .00175$, $f(.2) = .00349$, calculate $f(.15)$ approximately.

$$P(.15) = \frac{.15 - .2}{.1 - .2}.00175 + \frac{.15 - .1}{.2 - .1}.00349$$

$$= \frac{1}{2}.00175 + \frac{1}{2}.00349 = .00262$$

Example 3.

Given $f(x) = e^x$, $f(1) = 2.7183$, $f(1.10) = 3.0042$. Compute by linear interpolation $f(1.07)$.

$$P(1.07) = \frac{1.07 - 1.10}{1 - 1.10}2.7183 + \frac{1.07 - 1}{1.10 - 1}(3.0042)$$

$$= .3(2.7183) + (.7)(3.0042)$$

$$= 2.91843$$

Next we derive an error term for linear interpolation. We have the following theorem.

Theorem 4.1. Let $f(x) \epsilon c^2[x_0, x_1]$. Let x be a point in (x_0, x_1). Then there exists a point $\xi \epsilon (x_1, x_1)$ depending upon x such that

$$f(x) - P(x) = \frac{(x - x_0)(x - x_1)}{2} f''(\xi).$$

Form the function $F(t)$ as follows

$$F(t) = f(t) - P(t) - [f(x) - P(x)] \frac{(t - x_0)(t - x_1)}{(x - x_0)(x - x_1)}.$$

We now see that $F(x_0) = f(x_0) - P(x_0) - 0 = 0$

$$F(x_1) = f(x_1) - P(x_1) - 0 = 0,$$

and

$$F(x) = f(x) - P(x) - [f(x) - P(x)] \cdot 1 = 0.$$

Hence $F(t)$ is zero at x_0, x_1, and x. Now $F(t) \epsilon c^2[x_0, x_1]$ since $f(x) \epsilon c^2[x_0, x_1]$ and $P(t)$ and $\frac{(t - x_0)(t - x_1)}{(x - x_0)(x - x_1)}$ are polynomials hence in $c^2[x_0, x_1]$. In the interval $[x_0, x]$, $F'(\xi_1) = 0$ by Rolle's Theorem. Also in the interval $[x, x_1]$ $F'(\xi_2) = 0$ by Rolle's Theorem. Hence by Rolle's Theorem in the interval $[\xi_1, \xi_2]$ there exists a ξ such that $F''(\xi) = 0$. Computing $F''(t)$ we have

$$F''(t) = f''(t) - P''(t) - \frac{[f(x) - P(x)]}{(x - x_0)(x - x_1)} \frac{d^2}{dt^2}(t - x_0)(t - x_1)$$

Here $(t - x_0)(t - x_1) = t^2 - t(x_0 + x_1) + x_0x_1$;

hence $\frac{d^2}{dt^2}(t - x_0)(t - x_1) = 2.$

Now $P''(t) = 0$ since $P(t)$ is a linear function. Thus $F''(t)$ reduces to

$$F''(t) = f''(t) - \frac{[f(x) - P(x)] \cdot 2}{(x - x_0)(x - x_1)}.$$

Substituting ξ for t we have

$$F''(\xi) = 0 \text{ or } f''(\xi) = \frac{[f(x) - P(x)] \cdot 2}{(x - x_0)(x - x_1)}.$$

Simplifying we have

$$f(x) - P(x) = f''(\xi) \frac{(x - x_0)(x - x_1)}{2}.$$

The term $f''(\xi) \frac{(x - x_0)(x - x_1)}{2}$ is called the truncation error term of the approximation $P(x)$ to $f(x)$.

Example 4.

In example 2 bound the truncation error term. Here $f(x) = \sin(x)$, $f'(x) = \cos x$, $f''(x) = -\sin x$ so the truncation error is $f(x) - P(x) = -\sin(\xi) \frac{(.15 - .1)(.15 - .2)}{2}$. $|\sin(\xi)| \leq .00349$, so a bound is given by $|f(x) - P(x)| \leq \frac{.00349(.05)(.05)}{2} \leq .000004363$.

Example 5.

Bound the truncation error in Example 3. $f(x) = e^x$, $f''(x) = e^x$

$$|f(x) - P(x)| = e^{\xi} \frac{|(1 - 1.07)|\ |(1.1 - 1.07)|}{2}$$

$$\leqq \frac{3.0042 \cdot (.07)\ (.03)}{2} = .00315441$$

Example 6.

Obtain a bound that is somewhat easier to use. Let $M_2 = \max|f''(x)|$. So we have for $x \epsilon [x, x_1]$

$$|f(x) - P(x)| = |f''(\xi)| \frac{|(x - x_0)\ (x - x_1)|}{2} \leq M_2 \frac{|(x - x_0)(x - x_1)|}{2}$$

We ask, can we maximize $|(x - x_0)\ (x - x_1)|$ for $x \epsilon [x_0, x_1]$? It is a parabola that is zero at the endpoints. Taking the derivative of $Q(x) = (x - x_0)\ (x - x_1)$ we have $Q'(x) = x - x_0 + x - x_1 = 0$, or $x = \frac{x_0 + x_1}{2}$ for a maximum. Substituting this in we have

$|Q(x)| = \frac{(x_1 - x_0)^2}{4}$, or the bound

$$|f(x) - P(x)| \leq M_2 \frac{(x_1 - x_0)^2}{8}.$$

Exercises

1. Given $f(x) = e^x$, $f(0) = 1$, $f(.1) = 1.1052$, approximate $f(.07)$ by linear interpolation and bound the truncation error made.
2. Given $f(x) = \cos x$, $\cos(.1) = .99500$, $\cos(.2) = .98007$, approximate $f(.16)$ by linear interpolation and bound the error term.
3. Given $f(x) = \sin x$, $\sin(.6) = .56464$, $\sin(.7) = .64422$, approximate $f(.64)$ by linear interpolation and bound the error term.
4, Verify the identity

$$x = \frac{x - x_1}{x_0 - x_1} x_0 + \frac{(x - x_0)}{(x_1 - x_0)} x_1$$

by noting that $f(x) = x = P(x)$.

Section 4.2 Quadratic Interpolation. In this section we consider the following problem: a parabola $P(x) = Ax^2 + Bx + C$ to a set of data $f(x_0) = P(x_0)$, $f(x_1) = P(x_1)$, $f(x_2) = P(x_2)$. We first ask if we can do this, i.e., does there exist A, B, C such that the above conditions hold true.

We have the system of equations

$$P(x_0) = f(x_0) = Ax_0^2 + Bx_0 + C$$

$$P(x_1) = f(x_1) = Ax_1^2 + Bx_1 + C$$

$$P(x_2) = f(x_2) = Ax_2^2 + Bx_2 + C$$

The system has a solution provided that the determinant of coefficients does not vanish, i.e.,

$$\begin{vmatrix} x_0^2 & x_0 & 1 \\ x_1^2 & x_1 & 1 \\ x_2^2 & x_2 & 1 \end{vmatrix} \neq 0.$$

We evaluate the determinant by subtracting the first row from the second, and the first row from the third row, to obtain

$$\begin{vmatrix} x_0^2 & x_0 & 1 \\ x_1^2 - x_0^2 & x_1 - x_0 & 0 \\ x_2^2 - x_0^2 & x_2 - x_0 & 0 \end{vmatrix} = \begin{vmatrix} x_1^2 - x_0^2 & x_1 - x_0 \\ x_2^2 - x_0^2 & x_2 - x_0 \end{vmatrix}$$

Factoring out $x_1 - x_0$ from the first row and $x_2 - x_0$ from the second row we obtain

$$\begin{vmatrix} x_1^2 - x_0^2 & x_1 - x_0 \\ x_2^2 - x_0^2 & x_2 - x_0 \end{vmatrix} = (x_1 - x_0)(x_2 - x_0) \begin{vmatrix} x_1 + x_0 & 1 \\ x_2 + x_0 & 1 \end{vmatrix}$$

$$= (x_1 - x_0)(x_2 - x_0)(x_2 - x_1).$$

Thus the determinant is non-zero if $x_0 \neq x_1 \neq x_2$. If the points are distinct we can always solve for A, B, C.

Rather than solve the above system of equation we write down the result and verify that it works. We have

$$P(x) = \frac{(x - x_1)\,(x - x^2)}{(x_0 - x_1)(x_0 - x_2)}\,f(x_0) + \frac{(x - x_0)\,(x - x_2)}{(x_1 - x_0)\,(x_1 - x_2)}\,f(x_1)$$
$$+ \frac{(x - x_0)\,(x - x_1)}{(x_2 - x_0)\,(x_2 - x_1)}\,f(x_2)$$

We verify that $P(x_i) = f(x_i) \quad i = 0, 1, 2.$

$$P(x_0) = \frac{(x_0 - x_1)\,(x_0 - x_2)}{(x_0 - x_1)\,(x_0 - x_2)}\,f(x_0) + 0 \cdot f(x_1) + 0 \cdot f(x_2)$$
$$P(x_0) = 1 \cdot f(x_0) = f(x_0)$$

Similarly $P(x_1) = f(x_1)$, $P(x_2) = f(x_2)$.

In a manner similar to that for linear interpolation the truncation error term can be shown to be

$$\frac{(x - x_0)\,(x - x_1)\,(x - x_2)}{6}\,f'''(\xi).$$

Thus we have

$$f(x) - P(x) = \frac{(x - x_0)\,(x - x_1)\,(x - x_2)}{6}\,f'''\,(\xi).$$

Example 1.

Given sin (0) = 0, sin (.1) = .09983, sin (.2) = .19867, find sin (.05) by quadratic interpolation and bound the error term.

$$P(x) = \frac{(.05 - .1)\,(.05 - .2)}{(0 - .1)\,(0 - .2)} \cdot (0) + \frac{(.05 - 0)\,(.05 - .2)}{(.1 - 0)\,(.1 - .2)}\,(.09983)$$
$$+ \frac{(.05 - 0)\,(.05 - .1)}{(.2 - 0)\,(.2 - 1.)}\,(.19867)$$
$$= .75(.09983) + (-\,.125)\,(.19867)$$
$$= .07487 - .02483 = .05004.$$

Now $f''' = -\cos x$ so the error bound yields

$$|f(.05) - P(.05)| \leq \left| \frac{(.05 - 0)\,(.05 - .1)\,(.05 - .2)}{6} \cos \xi \right|$$
$$\leq .0000625$$

Here we used the the fact that $|\cos \xi| \leq 1$.

Example 2.

$f(x) = e^x$, $f(0) = 1.00$, $f(.2) = 1.2214$, $f(.4) = 1.4918$. Find an approximate value for $f(.1)$ and bound the error using quadratic interpolation.

$$P(.1) = \frac{(.1-.2)(.1-.4)}{(0-.2)(0-.4)} 1. + \frac{(.1-0)(.1-.4)}{(.2-0)(.2-.4)} (1.2214)$$
$$+ \frac{(.1-0)(.1-.2)}{(.4-1)(.4-.2)} (1.4918)$$

$$P(.1) = \frac{3}{8} + \frac{3}{4} (1.2214) + -\frac{1}{8}(1.4918)$$
$$= 1.10458 \cong 1.1046$$

$$f'''(x) = e^x$$

$$|P(.1) - f(.1)| = \left| \frac{(.1-.2)(.1-0)(.1-.4)}{2} e^\xi \right|$$
$$\leq .0015 \cdot 1.4918 \leq .00225.$$

Exercises

1. Given $f(x) = e^x$, $f(.1) = 1.1052$, $f(.2) = 1.2214$, $f(.3) = 1.3499$, find $f(.15)$ by quadratic interpolation and bound the error.
2. Given $f(x) = \cos x$, $f(0) = 1$, $f(.1) = .99500$, $f(.2) = .98007$, find $f(.16)$ by quadratic interpolation and bound the error.

Section 4.3 Lagrange Interpolation. In this section we fit a polynomial $P(x)$ to the data $x_0, x_1, x_2, \cdots, x_n$ where $P(x_i) = f(x_i)$ $i = 0, 1, 2, \cdots, n$.

We construct the Lagrangian interpolation polynomial as follows. We seek a representation of the form

$$P(x) = \sum_{i=0}^{n} \ell_i(x) f(x_i).$$

If we could choose the $\ell_i(x)$ so that $\ell_i(x_i) = 1$, $\ell_i(x_j) = 0$ $i \neq j$,

we would have $P(x_\ell) = \sum_{i=0}^{n} \ell_i(x_\ell) f(x_i) = f(x_\ell)$ and the interpolation problem is solved.

We choose $\ell_j(x)$ as follows.

$$\ell_j(x) = \frac{(x - x_0)(x - x_1)(x - x_2)\cdots(x - x_{j-1})(x - x_{j+1})\cdots(x - x_n)}{(x_j - x_0)(x_j - x_1)(x_j - x_2)\cdots(x_j - x_{j-1})(x_j - x_{j+1})\cdots(x_j - x_n)}$$

Then $\ell_j\,\epsilon(x_i) = \delta_{ij}$, $i, j = 1, 2, \cdots, n$.

Example 1.

Let $f(0) = 1, f(1) = 2, f(2) = 3, f(3) = 5$. Find the Lagrange interpolating polynomial through these points.

$$P(x) = \frac{(x-1)(x-2)(x-3)}{(0-1)(0-2)(0-3)}1 + \frac{(x-0)(x-2)(x-3)}{(1-0)(1-2)(1-3)}2 + \frac{(x-0)(x-1)(x-3)}{(2-0)(2-1)(2-3)}3 + \frac{(x-0)(x-1)(x-2)}{(3-0)(3-1)(3-2)}5$$

Example 2.

Let $f(x) = \sin x$, $f(.1) = .09983$, $f(.2) = .19867$, $f(.3) = .29552$, $f(.4) = .38942$.

Then $P(x)$ is given by

$$P(x) = \frac{(x-.2)(x-.3)(x-.4)}{(.1-.2)(.1-.3)(.1-.4)}.09983 + \frac{(x-.1)(x-.3)(x-.4)}{(.2-.1)(.2-.3)(.2-.4)}.19867 + \frac{(x-.1)(x-.2)(x-.4)}{(.3-.1)(.3-.2)(.3-.4)}.29552 + \frac{(x-.1)(x-.2)(x-.3)}{(.4-.1)(.4-.2)(.4-.3)}.38942$$

The following theorem can be established.

Theorem 4.2. Let $f(x)$ be $c^{n+1}[a, b]$. Let $x_0, x_1, x_2, \cdots, x_n$ and x lie in $[a, b]$. Then there exists a point ξ in the subinterval I of $[a, b]$ which contains $x_0, x_1, x_2, \cdots, x_n, x$ such that

$$f(x) = \sum_{j=0}^{n} f(x_j)\,\ell_j(x) + R_n(x)$$

where

$$R_n(x) = \frac{(x - x_0)(x - x_1)\cdots(x - x_n)}{(n+1)!} f^{n+1}(\xi)$$

Example 3.

Consider example 2 with $x = .15$. Compute a bound on the error term $R_n(x)$. We have $f(x) = \sin x$, $f'(x) = \cos x$, $f''(x) = -\sin x$, $f'''(x) = -\cos x$, $f^{iv}(x) = \sin x$.

$$R_n(.15) = \frac{(.15 - .1)(.15 - .2)(.15 - .3)(.15 - .4)}{24} f^{iv}(\xi)$$

$$|R_n(.15)[\leqq \frac{(.05)(.05)(.15)(.25)}{24} |\sin \xi| \leqq .00000016$$

We have used the fact that $|\sin(\xi)| \leq .38942 \leq .4$.

Example 4.

Let $f(x) = 1$, then $f^{n+1}(x) = 0$. Thus

$$1 = f(x) = \sum_{i=0}^{n} \ell_i(x) \cdot 1 = \sum_{i=0}^{n} \ell_i(x).$$

Example 5.

Let $P_*(x) = f(x)$, where $P_*(x)$ is a polynomial of degree less than or equal to n.
Then

$$P_*(x) = \sum_{i=0}^{n} \ell_j(x) P_*(x_j).$$

$$f^{n+1}(x) = P_*^{(n+1)} = 0$$

so

$$P_*(x) - P(x) = \sum_{i=0}^{n} \ell_j(x) P_*(x_j).$$

Exercises

1. Let $f(x) = \cos x$, $f(0) = 1$, $f(.1) = .99500$, $f(.2) = .98007$, $f(.3) = .95534$. Find $f(.19)$ and bound the truncation error made.
2. Let $f(x) = e^x$, $f(0) = 1$, $f(.1) = 1.1052$, $f(.2) = 1.2214$, $f(.3) = 1.3499$ $f(.4) = 1.4918$. Approximate $f(.22)$ by Lagrangian interpolation and bound the truncation error made.

3. Let $f(x) = \log x$, $f(1) = 0$, $f(1.1) = .09531$, $f(1.2) = .18232$, $f(1.3) = .26236$. Approximate $f(1.05)$ by Lagragian interpolation and bound the truncation error term.
4. Let $f(x) = \sinh(x)$, $f(0) = 0$, $f(.05) = .05002$, $f(.1) = .10017$. Find $f(.07)$ by Lagrangian interpolation and bound the error term.
5. Let $f(x) = \cosh x$, $f(0) = 1$, $f(.1) = 1.00500$, $f(.2) = 1.02007$. Approximate $f(.15)$ by Lagrangian interpolation and bound the truncation error.

Section 4.4 The Newton Divided Difference Formula. Consider the function $f(x)$ which has values at the discrete set of distant points $x_0, x_1, x_2, \cdots, x_n$. We define the first divided difference $[x_0\, x_1]$ as

$$[x_0\, x_1] = \frac{f(x_0) - f(x_1)}{x_0 - x_1} = \frac{f(x_1) - f(x_0)}{x_1 - x_0} = [x_1\, x_0].$$

The second divided difference is defined as

$$[x_0\, x_1\, x_2] = \frac{[x_0\, x_1] - [x_1\, x_2]}{x_0 - x_2}$$

Proceeding in this way we have

$$[x_0\, x_1 \cdots x_n] = \frac{[x_0\, x_1 \cdots x_{n-1}] - [x_1\, x_2 \cdots x_n]}{x_0 - x_n}$$

A table of the divided differences is given as follows:

$$\begin{array}{llllll}
x_0 & f(x_0) & & & & \\
 & & [x_0\, x_1] & & & \\
x_1 & f(x_1) & & [x_0\, x_1\, x_2] & & \\
 & & [x_1\, x_2] & & [x_0\, x_1\, x_2\, x_3] & \\
x_2 & f(x_2) & & [x_1\, x_2\, x_3] & & [x_0\, x_1\, x_2\, x_3\, x_4] \\
 & & [x_2\, x_3] & & [x_1\, x_2\, x_3\, x_4] & \cdot \\
x_3 & f(x_3) & & [x_2\, x_3\, x_4] & \cdot & \cdot \\
 & & [x_3\, x_4] & \cdot & \cdot & \cdot \\
 & & \cdot & \cdot & \cdot & \\
x_4 & f(x_4) & \cdot & \cdot & [x_{n-3}\, x_{n-2}\, x_{n-1}\, x_n] & \\
\cdot & & \cdot & [x_{n-2}\, x_{n-1}\, x_n] & & \\
\cdot & & & & & \\
\cdot & & [x_{n-1}\, x_n] & & & \\
x_n & & & & &
\end{array}$$

We write $[x\,x_1] = \dfrac{f(x) - f(x_1)}{x - x_1}$ as

$$f(x) = f(x_1) + (x - x_1)\,[x\,x_1] \tag{4.1}$$

and then

$$[x\,x_1\,x_2] = \frac{[x\,x_1] - [x_1\,x_2]}{x - x_2}$$

so $[x\,x_1] = [x_1\,x_2] + (x - x_2)\,[x\,x_1\,x_2]$. Substituting this in (4.1) we have

$$f(x) = f(x_1) + (x - x_1)\,[x_1\,x_2] + (x - x_1)(x - x_2)\,[x\,x_1\,x_2]. \tag{4.2}$$

Now

$$[x\,x_1\,x_2\,x_3] = \frac{[x\,x_1\,x_2] - [x_1\,x_2\,x_3]}{x - x_3}$$

so

$$[x\,x_1\,x_2] = [x_1\,x_2\,x_3] + (x - x_3)\,[x\,x_1\,x_2\,x_3]$$

Hence, substituting in 4.2 we have

$$\begin{aligned} f(x) = f(x_1) &+ (x - x_1)\,[x_1\,x_2] + (x - x_1)(x - x_2)\,[x_1\,x_2\,x_3] \\ &+ (x - x_1)(x - x_2)(x - x_3)\,[x\,x_1\,x_2\,x_3]. \end{aligned} \tag{4.3}$$

We continue this process of substitution and obtain finally

$$\begin{aligned} f(x) = f(x_1) &+ (x - x_1)\,[x_1\,x_2] + (x - x_1)(x - x_2)\,[x_1\,x_2\,x_3] \\ &+ \cdots + (x - x_1)(x - x_2)\cdots(x - x_{n-1})[x_1\,x_2\cdots x_n] \\ &+ (x - x_1)(x - x_2)\cdots(x - x_n)\,[x\,x_1\,x_2\cdots x_n] \end{aligned}$$

The term $(x - x_1)(x - x_2)\cdots(x - x_n)\,[x\,x_1\,x_2\cdots x_n]$ is called the truncation error term. Now it can be shown that

$$[x\,x_1\,x_2\cdots x_n] = \frac{f^n(\xi)}{n!}$$

where ξ is in an interval containing the points $x, x_1, x_2, \cdots, x_n$. So the truncation error term can be represented as

$$R_n(x) = (x - x_1)(x - x_2)\cdots(x - x_n)\,\frac{f^n(\xi)}{n!}.$$

Then

$$\begin{aligned} P(x) = f(x_1) &+ (x - x_1)\,[x_1\,x_2] + (x - x_1)(x - x_2)\,[x_1\,x_2\,x_3] \\ &+ \cdots + (x - x_1)(x - x_2)\cdots(x - x_{n-1})\,[x_1\,x_2\cdots x_n] \end{aligned}$$

is a polynomial passing through the points $f(x_1), f(x_2), \cdots, f(x_n)$.

Example 1.

Given $x_0 = 0$, $x_1 = .2$, $x_2 = .3$, $f(x) = \sin x$, interpolate using the above formula for f(.23).

x	f(x)		
$x_0 = 0$	.00000		
		.99335	
$x_1 = .2$	.19867		$-.08281$
		.96850	
$x_2 = .3$	.29552		

$$f(.23) \approx .00000 + (.23)(.99335) + (.23)(.03)(-.08281)$$
$$= .217908$$

Example 2.

Given $x_0 = 0$, $x_1 = .1$, $x_2 = .2$, $f(x) = e^x$, interpolate using the above formula for f(.15).

x	f(x)		
$x_0 = 0$	1.0000		
		1.0520	
$x_1 = .1$	1.1052		.0500
		1.1620	
$x_2 = .2$	1.2214		

$$f(.15) = 1.0000 + (.15)(1.0520) + (.15)(.05)(.0500)$$
$$= 1.158175 \cong 1.1582$$

Exercises

1. Given $f(x) = e^x$, $x_0 = 0$, $x_1 = .1$, $x_2 = .2$, $x_3 = .3$, interpolate for f(.15) using the above interpolating polynomial.
2. Given $f(x) = \cos x$, $x_0 = 0$, $x_1 = .1$, $x_2 = .2$, interpolate for cos (.23) using the above interpolating polynomial.
3. Given $f(x) = x^2$ write out $[x_0\, x_1]$, $[x_0\, x_1\, x_2]$.
4. By using the definition $[x\, x_1] = \dfrac{f(x)}{x - x_1} + \dfrac{f(x_1)}{x_1 - x}$, derive the identity

$$[x\, x_1\, x_2] = \frac{f(x)}{(x - x_1)(x - x_2)} + \frac{f(x_1)}{(x_1 - x)(x_1 - x_2)} + \frac{f(x_2)}{(x_2 - x)(x_2 - x_1)}.$$

Section 4.5 Aitken's Method of Interpolation. A method of interpolation which is particularly amenable to digital computers is Aitken's method of iterated linear interpolation.

Linear interpolation can be viewed as

$$y_{0,1}(x) = \frac{1}{x_1 - x_0}\begin{vmatrix} y_0(x) & x_0 - x \\ y_1(x) & x_1 - x \end{vmatrix}$$

where the interpolating polynomials of degree zero are

$$y_0(x) = f(x_0),\ y_1(x) = f(x_1).$$

In the same way, quadratic interpolation can be viewed as interpolation over two independent linear interpolation polynomials

$$y_{0,1,2}(x) = \frac{1}{x_2 - x_0}\begin{vmatrix} y_{0,1}(x) & x_0 - x \\ y_{1,2}(x) & x_2 - x \end{vmatrix}$$

$$= \frac{1}{x_2 - x_1}\begin{vmatrix} y_{0,1}(x) & x_1 - x \\ y_{0,2}(x) & x_2 - x \end{vmatrix}$$

We verify this directly. We note that the right-hand side is a polynomial of degree two

$$y_{0,1,2}(x_0) = \frac{1}{x_2 - x_0}\begin{vmatrix} f(x_0) & 0 \\ y_{0,2}(x_0) & x_2 - x_0 \end{vmatrix} = f(x_0)$$

Similarly,

$$y_{0,1,2}(x_1) = f(x_1),\quad y_{0,1,2}(x_2) = f(x_2).$$

Similarly, we can build up higher degree interpolation polynomials. In Aitken's scheme we have

x_0	y_0					$x_0 - \bar{x}$
x_1	y_1	$y_{0,1}$				$x_1 - \bar{x}$
x_2	y_2	$y_{0,2}$	$y_{0,1,2}$			$x_2 - \bar{x}$
x_3	y_3	$y_{0,3}$	$y_{0,1,3}$	$y_{0,1,2,3}$		$x_3 - \bar{x}$
x_4	y_4	$y_{0,4}$	$y_{0,1,4}$	$y_{0,1,2,4}$	$y_{0,1,2,3,4}$	$x_4 - \bar{x}$

so that

$$y_{0,1,2} = \frac{1}{x_2 - x_1}\begin{vmatrix} y_{0,1} & x_1 - \bar{x} \\ y_{0,3} & x_3 - \bar{x} \end{vmatrix}$$

$$y_{0,1,2,3,4} = \frac{1}{x_4 - x_3}\begin{vmatrix} y_{0,1,2,3} & x_3 - \bar{x} \\ y_{0,1,2,4} & x_4 - \bar{x} \end{vmatrix}$$

An example is: Compute $f(x) = x^3$ at $x = 3$, from $x_0 = 0$, $x_1 = 1$, $x_2 = 2$, $x_3 = 4$.

					$x_i - \bar{x}$
0	0				-3
1	1	3			-2
2	8	12	21		-1
4	64	48	33	27	1

$$y_{0,1} = \frac{1}{1-0}\begin{vmatrix} 0 & -3 \\ 1 & -2 \end{vmatrix} = 3$$

$$y_{0,2} = \frac{1}{2-0}\begin{vmatrix} 0 & -3 \\ 8 & -1 \end{vmatrix} = 12$$

$$y_{0,3} = \frac{1}{4-0}\begin{vmatrix} 0 & -3 \\ 64 & 1 \end{vmatrix} = 48$$

$$y_{0,1,2} = \frac{1}{2-1}\begin{vmatrix} 3 & -2 \\ 12 & -1 \end{vmatrix} = 21$$

$$y_{0,1,3} = \frac{1}{4-1}\begin{vmatrix} 3 & -2 \\ 48 & 1 \end{vmatrix} = \frac{1}{3}\cdot 99 = 33$$

$$y_{0,1,2,3} = \frac{1}{4-2}\begin{vmatrix} 21 & -1 \\ 33 & 1 \end{vmatrix} = \frac{1}{2}\cdot 54 = 27$$

A Fortran program for computing the interpolated value using Aitken's method follows:

```
      DIMENSION X(20), XS(20), Y(20, 20)
      READ (5, 1) XX, N
 1    FORMAT (F15.7, I2)
      DO 2I = 1, N
 2    READ (5, 3) X(I), Y(I, 1)
 3    FORMAT (2F 15.7)
      DO 4I = 1, N
 4    XS(I) = X(I) - XX
      DO 5I = 2, N
      DO 5J = 2, I
      Y(I, J) = Y(J - 1, J - 1) * XS(I) - Y(I, J - 1)
                                  * XS(J - 1)
      Y(I, J) = Y(I, J)/(X(I) - X(J - 1))
 5    CONTINUE
      DO 6I = 1, N
      WRITE (6, 7) X(I), (Y(I, J), J = 1, I)
```

```
7    FORMAT (F15.7, 7(E15.7),/, 2X, 7(E15.7), 1, 4X,
     6(E15.7))
6    CONTINUE
     STOP
     END
```

Exercises

1. Given $f(x) = e^x$, $x_0 = 0$, $x_1 = .1$, $x_2 = .2$, $x_3 = .3$, compute an interpolated value for f(.4) using Aitken's method.
2. Given $f(x) = \sin x$, $x_0 = 0$, $x_1 = .1$, $x_2 = .2$, $x_3 = .3$, compute an interpolated value for f(.15) using Aitken's method.
3. Write a Fortran program to compute f(.05) where $f(x) = \cos(x)$, $x_0 = 0$, $x_1 = .1$, $x_2 = .2$, $x_3 = .3$, $x_4 = .4$.

Chapter 5

NUMERICAL INTEGRATION

5.1 Introduction. In calculus we are taught to integrate certain functions explicitly. In practice, however, the integrals may be too complicated to integrate explicitly or it my be that no formula exists for the integral such as $\int_0^x e^{t^2}\,dt$. Thus we must take recourse to numerical methods.

In this chapter we shall give some of the more popular numerical methods for numerical integration.

Section 5.2 The Trapezoidal Rule. If we linearly interpolate f(x) between x_0 and $x_1 = x_0 + h$, we obtain

$$f(x) = \frac{(x - x_1)}{(x_0 - x_1)} f(x_0) + \frac{(x - x_0)}{(x_1 - x_0)} f(x_1) + \frac{(x - x_0)(x - x_1)}{2} f''(\xi) \tag{1}$$

Integrating equation (1) between x_0 and x_1 we have

$$\int_{x_0}^{x_1} f(x)dx = \frac{f(x_0)}{-h}\int_{x_0}^{x_1} (x - x_1)dx + \frac{f(x_1)}{h}\int_{x_0}^{x_1} (x - x_0)dx$$
$$+ \int_{x_0}^{x_1} \frac{(x - x_0)(x - x_1)}{2} f''(\xi)dx \tag{2}$$

Integrating

$$\int_{x_0}^{x_1} (x - x_1)dx = \left.\frac{(x - x_1)^2}{2}\right|_{x_0}^{x_1} = -\frac{h^2}{2}$$

and

$$\int_{x_0}^{x_1} (x - x_0)dx = \left.\frac{(x - x_0)^2}{2}\right|_{x_0}^{x_1} = \frac{h^2}{2}$$

we have

$$\int_{x_0}^{x_1} f(x)dx = \frac{f(x_0)}{-h} \cdot \frac{-h^2}{2} + \frac{f(x_1)h^2}{h2}$$

$$+ \int_{x_0}^{x_1} \frac{(x - x_0)(x - x_1)}{2} f''(\xi)dx \qquad (3)$$

Simplifying we have

$$\int_{x_0}^{x_1} f(x)dx = \frac{h}{2}[f(x_0) + f(x_1)]$$

$$+ \int_{x_0}^{x_1} \frac{(x - x_0)(x - x_1)}{2} f''(\xi)dx \qquad (4)$$

The approximation $\int_{x_0}^{x_1} f(x)dx \approx \frac{h}{2}[f(x_0) + f(x_1)]$ is called the trapezoidal rule.

The term $\int_{x_0}^{x_1} \frac{(x - x_0)(x - x_1)}{2} f''(\xi)dx$ is called the truncation error term. We shall now simplify this term.

We note that $\frac{(x - x_0)(x - x_1)}{2} \leq 0$ for $x \epsilon [x_0, x_1]$. This allows us to use the second mean value theorems for integrals with $g(x) = (x - x_0)(x - x_1)/2$ and $F(x) = f''(\xi)$. We obtain

$$\int_{x_0}^{x_1} \frac{(x - x_0)(x - x_1)}{2} f''(\xi)dx =$$

$$f''(\xi_1) \int_{x_0}^{x_1} \frac{(x - x_0)(x - x_1)}{2} dx. \qquad (5)$$

To evaluate the integral on the right-hand side of equation (5) we let $x = x_0 + sh$. Then $x - x_1 = x_0 + sh - (x_0 + h) = (s - 1)h$, $dx = h\ ds$, $x - x_0 = sh$. When $x = x_0$, $s = 0$, when $x = x_1 = x_0 + h$, $s = 1$. So we obtain

$$\frac{h^3}{2}\int_0^1 s(s - 1)ds = \frac{h^3}{2}\left[\frac{s^3}{3} - \frac{s^2}{2}\right]_0^1 = -\frac{h^3}{12}$$

Thus equation (4) now reads

$$\int_{x_0}^{x_1} f(x)dx = \frac{h}{2}[(fx_0) + f(x_1)] - \frac{h^3}{12}f''(\xi). \tag{6}$$

This is the trapezoidal rule with truncation error term.

Example 1.

Evaluate $\int_0^{.1} e^x\, dx$ using the trapezoidal rule and bound the error made.

$$\int_0^{.1} e^x\, dx \approx \frac{.1}{2}[e^0 + e^{.1}] = \frac{.1}{2}[1 + 1.1052]$$

$$= .10526 = .1053 \text{ to } 4 \text{ D}.$$

The exact value to 4D is

$$\int_0^{.1} e^x\, dx = e^{.1} - e^0 = 1.1052 - 1 = .1052.$$

Now $f(x) = e^x$, so $f''(x) = e^x$; hence the error is $-\frac{(.1)^3}{12}e^{\xi}$. A bound on this is $\frac{(.1)^3}{12}e^{.1} = \frac{(.1)^3}{12}(1.1052) = .0000921.$

Example 2.

Evaluate the integral $\int_{.1}^{.2} \sin x\, dx$, using the trapezoidal rule.

$$\int_{.1}^{.2} \sin x\, dx = \frac{.1}{2}[\sin .1 + \sin .2]$$

$$= \frac{.1}{2}[.09983 + .19867]$$

$$= .014925 \approx .01493$$

whereas the exact value is

$$\int_{.1}^{.2} \sin(x)dx = -\cos(x)\Big|_{.1}^{.2}$$

$$= .99500 - .98007 = .01493.$$

Consider $\int_0^1 e^x\,dx$. By the trapezoidal rule we obtain

$$\int_0^1 e^x\,dx \approx \frac{1}{2}[e^0 + e^1] = 1.85915$$

whereas the exact value is $\int_0^1 e^x\,dx = e^x\Big|_0^1 = 1.7183$. The large error encountered here is due to the fact that $h = 1$ is too large. We ask, could we modify the trapezoidal rule to take care of such cases? Below we shall do this.

Suppose we have $\int_a^b f(x)dx$. We let $x_0 = a$, $(b - a)/n = h$, $x_n = b$ where n is an integer. Then if we integrate from x_0 to $x_0 + h$, $x_0 + h$ to $x_0 + 2h$, etc., $x_0 + (n - 1)h$ to $x_0 + nh$, we have

$$\int_{x_0}^{x_0+h} f(x)dx = \frac{h}{2}[f(x_0) + f(x_0 + h)] - \frac{h^3}{12}f''(\xi_1)$$

$$\int_{x_0+h}^{x_0+2h} f(x)dx = \frac{h}{2}[f(x_0 + h) + f(x_0 + 2h)] - \frac{h^3}{12}f''(\xi_2)$$

$$\vdots$$

$$\int_{x_0+(n-1)h}^{x_0+(nh)} f(x)dx = \frac{h}{2}[f(x_{n-1}) + f(x_n)] - \frac{h^3}{12}f''(\xi_n)$$

Adding up the left-hand side we have

$$\int_{x_0}^{x_0+h} f(x)dx + \int_{x_0+h}^{x_0+2h} f(x)dx + \cdots + \int_{x_0+(n-1)h}^{(x_0+nh)} f(x)dx = \int_a^b f(x)dx$$

Adding up the right-hand side we have

$$h\left[\frac{f(x_0)}{2} + f(x_1) + f(x_2) + \cdots + f(x_{n-1}) + \frac{f(x_n)}{2}\right] - \frac{h^3}{12}[f''(\xi_1) + f''(\xi_2) + \cdots f''(\xi_n)]$$

Thus we have

$$\int_a^b f(x)dx = h\left[\frac{f(x_0)}{2} + f(x_1) + f(x_2) + \cdots + f(x_{n-1}) + \frac{f(x_n)}{2}\right] - \frac{h^3}{12}[f''(\xi_1) + f''(\xi_2) + \cdots + f''(\xi_n)].$$

Example 1.

Integrate $\int_0^{.2} e^x\, dx$ with $n = 2$, $h = (.2 - 0)/2 = .1$.

$$\int_0^{.2} e^x\, dx \approx .1\left[\frac{e^0}{2} + e^{.1} + \frac{e^{.2}}{2}\right]$$

$$= \frac{.1}{2}\left[\frac{1}{2} + 1.1052 + \frac{1.2214}{2}\right]$$

$$= .22159 \approx .2216$$

The error is $-\frac{(.1)^3}{12}[f''(\xi_1) + f''(\xi_2)]$. Since $f(x) = e^x$, $f''(x) = e^x$

and $\left|\frac{(.1)^3}{12}[f''(\xi_1) + f''(\xi_2)\right| \leq \frac{(.1)^3}{12} \cdot 2 \cdot e^{.2} < .00021.$

The truncation error term can be bounded in the following manner

$$\left|\frac{h^3}{12}(f''(\xi_1) + f''(\xi_2) + \cdots + f''(\xi_n))\right| \leq \frac{h^3}{12} n \cdot \max_{x \in [a, b]} |f''(x)|.$$

Example 2.

Apply the error bound to $\int_0^1 e^x\, dx$ with $h = .1$, hence $n = (1 - 0)/.1 = 10$.

$$f''(x) = e^x$$

so

$$\frac{h^3}{12} n \max_{x \in [0, 1]} | f''(x) | = \frac{(.1)^3}{12} \cdot 10 \cdot e.$$

$$\frac{(.1)^3}{12} \cdot 10 \cdot e < .001018.$$

Write a Fortran program to compute $\int_0^1 e^x \sin(x)dx$, $h = .1$, by the trapezoidal rule. Here $N = 1/.1 = 10$.

```
      F(X) = EXP(X)*SIN(X)
      SUM = F(0.)/2. + F(1.)/2.
      H = .1
      X = 0.1
      N = 9
      DO 1 I = 1, N
      SUM = SUM + F(X)
      X = X + H
    1 CONTINUE
      APPROX = H*SUM
      WRITE (6, 10) H, APPROX
   10 FORMAT (F10.2, E20.8)
      STOP
      END
```

Exercises

1. Evaluate $\int_{.1}^{.2} e^x\, dx$ by the trapezoidal rule $h = .1$ and bound the truncation error term.

2. Evaluate $\int_0^{.1} x^3\, dx$ by the trapezoidal rule $h = .1$ and bound the error term.

3. Evaluate $\int_{x_0}^{x_0 + h} (A + Bx)dx$ by the trapezoidal rule. Calculate the truncation error term.

4. Write a Fortran program to evaluate $\int_0^1 e^x\,dx$ with $h = .05$.

   Bound the truncation error term.

5. Write a Fortran program to evaluate $\int_0^1 \sin x\,dx$ with $h = .05$.

   Bound the truncation error term.

6. How large must n be in order that the truncation error be less than 10^{-4} for $\int_0^1 e^x\,dx$.

Section 5.3 Simpson's Rule. In the derivation of the trapezoidal rule we saw that the error was order of accuracy h^3. In this section we derive a method of numerical integration which will be of a higher order accuracy than the trapezoidal rule.

We approximate f(x) by a parabola passing through the points x_0, $x_0 + h = x_1$, $x_0 + 2h = x_2$.

$$f(x) = \frac{(x - x_1)(x - x_2)}{(x_0 - x_1)(x_0 - x_2)} f(x_0) + \frac{(x - x_0)(x - x_2)}{(x_1 - x_0)(x_1 - x_2)} f(x_1)$$

$$+ \frac{(x - x_0)(x - x_1)}{(x_2 - x_0)(x_2 - x_1)} f(x_2) + \frac{(x - x_0)(x - x_1)(x - x_2)}{6} f'''(\xi) \qquad (1)$$

We integrate equation (1) from x_0 to x_2 to obtain

$$\int_{x_0}^{x_2} f(x)dx = \frac{f(x_0)}{-h \cdot -2h} \int_{x_0}^{x_2} (x - x_1)(x - x_2)dx + \frac{f(x_1)}{h \cdot -h} \cdot \int_{x_0}^{x_2} (x - x_0)(x - x_2)dx$$

$$+ \frac{f(x_2)}{2h \cdot h} \int_{x_0}^{x_2} (x - x_0)(x - x_1)dx$$

$$+ \int_{x_0}^{x_2} \frac{(x - x_0)(x - x_1)(x - x_2)}{6} f'''(\xi)dx \qquad (2)$$

Letting $x = x_0 + sh$, we obtain $s = 0$ for the limit of integrals $x_0 = x_0 + sh$ and $s = 2$ for $x_2 = x_0 + 2h$. Also $dx = hds$. Thus we have

$$\int_{x_0}^{x_2} f(x)dx = \frac{f(x_0)}{2h^2}\int_0^2 (s-1)(s-2)h^3 ds + \frac{f(x_1)}{-h^2}\int_0^2 s(s-2)h^3ds$$

$$+ \frac{f(x_2)}{2h^2}\int_0^2 s(s-1)h^3 ds$$

$$+ \int_{x_0}^{x_2} \frac{(x-x_0)(x-x_1)(x-x_2)}{6} f'''(\xi)dx. \qquad (3)$$

Evaluating the integrals on the right-hand side of equation (3) we have

$$\int_0^2 (s-1)(s-2)ds = \int_0^2 (s^2 - 3s + 2)ds$$

$$= \left[\frac{s^3}{3} - \frac{3s^2}{2} + 2s\right]_0^2 = \frac{2}{3}$$

$$\int_0^2 s(s-2)ds = \int_0^2 (s^2 - 2s)ds = \left[\frac{s^3}{3} - \frac{2s^2}{2}\right]_0^2 = -\frac{4}{3}$$

$$\int_0^2 s(s-1)ds = \int_0^2 (s^2 - s)ds = \left[\frac{s^3}{3} - \frac{s^2}{2}\right]_0^2 = \frac{2}{3}$$

Thus equation (3) simplifies to read

$$\int_{x_0}^{x_2} f(x)dx = \frac{h}{3}\,[f(x_0) + 4f(x_1) + f(x_2)] + \text{Error} \qquad (4)$$

This is Simpson's rule. The error term can be simplified to yield $-\frac{h^5}{90} f^{iv}(\xi)$, $\xi \in [x_0, x_2]$. We do not give a derivation here. See Hildebrand [ ].

Example 1.

Evaluate $\int_0^{.2} \sin x\, dx$ by Simpon's rule and bound the truncation error.

Here h = .1; thus Simpson's rule yields

$$\int_0^{.2} \sin x \, dx \approx \frac{.1}{3} [\sin 0 + 4 \sin (.1) + \sin (.2)]$$

$$= \frac{.1}{3} [0 + 4(.09983) + .19867]$$

$$= .019933 \approx .01993$$

Since $f(x) = \sin x$, $f'(x) = \cos x$, $f''(x) = -\sin x$, $f'''(x) = -\cos x$, $f^{iv}(x) = \sin x$.

The truncation error term is

$$-\frac{h^5}{90} f^{iv}(\xi) = -\frac{(.1)^5}{90} \sin(\xi)$$

$$\left| -\frac{(.1)^5}{90} \sin(\xi) \right| < \frac{(.1)^5}{90} \cdot .19867 < 3.10^{-8}.$$

Example 2.

Evaluate $\int_{.2}^{.4} e^x \, dx$ by Simpson's rule and bound the truncation error made.

Here h = .1, so

$$\int_{.2}^{.4} e^x \, dx \approx \frac{.1}{3} [e^{.2} + 4e^{.3} + e^{.4}]$$

$$= \frac{.1}{3} [1.2214 + 4 \cdot (1.3499) + 1.4918]$$

$$\approx .27043 \approx .2704.$$

Here $f(x) = e^x$, $f^{iv}(x) = e^x$

$$\left| \frac{h^5}{90} f^{iv}(\xi) \right| = \frac{(.1)^5}{90} e^{\xi} < \frac{(.1)^5}{90} \cdot (1.5)$$

$$< (1.7) \cdot 10^{-7}.$$

Let $h = (b - a)/2m$; then Simpson's rule for 2m subintervals yields

$$\int_a^b f(x)dx = \frac{h}{3} [f(a) + f(b)] + 2 \sum_{i=1}^{m-1} f(x_{2i})$$

$$+ \sum_{i=1}^{m} f(x_{2i-1})] - m \frac{h^5}{90} f^{iv}(\xi).$$

Here $x_i = a + ih$, $i = 0, 1, 2, \cdots, 2m$.

Apply Simpson's rule to $\int_0^2 e^x \, dx$ with $m = 4$. Here

$h = (2 - 0)/2{\cdot}4 = \frac{1}{4}$.

So Simpson's rule yields

$$= \frac{1}{4} \cdot \frac{1}{3} [e^0 + 4e^{.25} + 2e^{.5} + 4e^{.75} + 2e^{1.} + 4e^{1.25} + 2e^{1.5} + 4e^{1.75} + e^{2.}]$$

The truncation error is bounded by

$$\left| \frac{-4e^{\xi}}{4^5 \cdot 90} \right| < \frac{e^2}{4^4 \cdot 90} \approx \frac{7.3891}{256{\cdot}90} < .00033.$$

Exercises

1. Evaluate $\int_{.1}^{.3} e^x \, dx$ by Simpson's rule with $h = .1$ and bound the truncation error.
2. Integrate $\int_0^1 x^3 \, dx$ by Simpson's rule and obtain an expression for the truncation error.
3. Evaluate $\int_0^{.1} \sin x \, dx$ by Simpson's rule with $h = .05$ and bound the truncation error.
4. Evaluate $\int_0^1 e^x \, dx$; use $m = 2$ by Simpson's rule.
5. Write a Fortran program to evaluate $\int_0^2 \sin x \, dx$, using Simpson's rule with $m = 10$, $m = 20$.

6. Integrate $\int_0^1 \frac{dx}{1+x}$ by Simpson's rule with $h = .5$. Compare your result with the exact result.

7. Evaluate $\int_1^2 \frac{\sin x}{x}\, dx$ by Simpson's rule with $h = .5$.

Section 5.4 Other Numerical Integration Formulae. Consider

$$\int_{x_0}^{x_0+h} f(x)dx.$$

Expand $f(x)$ in a Taylor's series about $f(x_0)$. Hence $f(x) = f(x_0) + (x - x_0)\, f'(\xi)$.

Integrating $f(x)$ from $\int_{x_0}^{x_1} f(x)dx$. We obtain

$$\int_{x_0}^{x_0+h} f(x)dx = \int_{x_0}^{x_0+h} [f(x_0) + (x - x_0)f'(\xi)]dx$$

$$= hf(x_0) + \int_{x_0}^{x_0+h} (x - x_0)\, f'(\xi)dx.$$

Since $x - x_0 \geq 0$, $x \in [x_0, x_0 + h]$ we can apply the second mean value theorem for integrals with $g(x) = x - x_0$, $F(x) = f'(\xi)$ to obtain

$$\int_{x_0}^{x_0+h} f(x)dx = hf(x_0) + f'(\xi_1) \int_{x_0}^{x_0+h} (x - x_0)dx. \quad (1)$$

Since

$$\int_{x_0}^{x_0+h} (x - x_0)dx = \frac{(x - x_0)^2}{2} \Bigg|_{x_0}^{x_0+h} = \frac{h^2}{2},$$

we have equation (1) which reads

$$\int_{x_0}^{x_0+h} f(x)dx = h\,f(x_0) + \frac{h^2}{2} f'(\xi_1),\ \xi_1 \in [x_0, x_0 + h] \quad (2)$$

It is easily verified that

$$\int_{x_0}^{x_0 + nh} f(x)dx = h[f(x_0) + f(x_1) + \cdots + f(x_{n-1})] + \frac{h^2}{2}[f'(\xi_1) + \cdots + f'(\xi_n)] \quad (3)$$

A bound on the error is given by

$$\frac{nh^2}{2} \max_{x \epsilon [x_0,\, x_0 + nh]} |f'(x)|$$

Due to the fact that the error is $\frac{nh^2}{2} \max_{x \epsilon [x_0,\, x_0 + nh]} |f'(x)|$, equation (3) is usually not preferable to the trapezoidal rule. However, if $f(x) \epsilon c^1 [a, b]$ and $f(x)c^2 [a, b]$, it would be preferable to use such a numerical integration scheme.

Next we show the integration formula

$$\int_{x_0 - h}^{x_0 + h} f(x)dx = 2h\, f_0 + \frac{h^3}{3} f''(\xi_1).$$

This formula is called the midpoint rule. If we expand f(x) in a Taylor's series about x, we obtain

$$f(x) = f(x_0) + (x - x_0)\, f'(x_0) + \frac{(x - x_0)^2}{2} f''(\xi)$$

Integrating f(x) between $x_0 - h$ and $x_0 + h$ we obtain

$$\int_{x_0 - h}^{x_0 + h} f(x)dx = x\, f(x_0)\Big|_{x_0 - h}^{x_0 + h} + \frac{(x - x_0)^2}{2} f'(x)\Big|_{x_0 - h}^{x_0 + h} + \int_{x_0 - h}^{x_0 + h} \frac{(x - x_0)^2}{2} f''(\xi)dx$$

$$\int_{x_0 - h}^{x_0 + h} f(x)dx = 2h\, f(x_0) + 0 \cdot f'(x_0) + \int_{x_0 - h}^{x_0 + h} \frac{(x - x_0)^2}{2} f''(\xi)dx$$

The error term can be treated by using the second mean value theorem since

$$\frac{(x - x_0)^2}{2} \geq 0 \quad x \epsilon [x_0 - h, x_0 + h].$$

Thus we have

$$\int_{x_0 - h}^{x_0 + h} \frac{(x - x_0)^2 f''(\xi)}{2} dx$$

$$= f''(\xi_1) \int_{x_0 - h}^{x_0 + h} \frac{(x - x_0)^2}{2} dx$$

$$= f''(\xi_1) \frac{(x - x_0)^3}{6} \Big|_{x - h}^{x_0 + h} = \frac{h^3}{3} f''(\xi_1)$$

Hence the desired formula

$$\int_{x_0 - h}^{x_0 + h} f(x)dx = 2hf(x_0) + \frac{h^2}{3} f''(\xi). \tag{4}$$

Next we derive the formula

$$\int_{x_0 - h}^{x_0 + h} f(x)dx = \frac{3h}{2} (f(x_0) + f(x_0 + h)) + \frac{3h^3}{4} f''(\xi) \tag{5}$$

We fit f(x) to a linear interpolating polynomial through x_0 and $x_1 = x_0 + h$ to obtain

$$f(x) = \frac{x - x_1}{x_0 - x_1} f(x_0) + \frac{(x - x_0)}{(x_1 - x_0)} f(x_1) + \frac{(x - x_0)(x - x_1)}{2} f''(\xi) \tag{6}$$

Integrating equation (6) from $x_0 - h$ to $x_0 + 2h$ we have, letting $x = x_0 + sh$, $s = -1$, $s = 2$, the limits of integration.

$$\int_{x_0 - h}^{x_0 + 2h} f(x)dx = h f(x_0) \int_{-1}^{2} \frac{(s - 1)h}{-h} ds$$

$$+ \int_{-1}^{2} \frac{sh}{h} h\, ds \cdot f(x_1) + \text{Error}$$

Simplifying, we have

$$\int_{x_0 - h}^{x_0 + 2h} f(x)dx = -h f(x_0) \cdot \frac{(s - 1)^2}{2} \Big|_{-1}^{2} + h f(x_1) \frac{s^2}{2} \Big|_{-1}^{2} + \text{Error}$$

$$= \frac{3}{2} h \,[f(x_0) + f(x_1)] + \text{Error}$$

Next we derive an integration formula of the form $\int_0^1 F(x)dx = W_0 F(x_0)$ by requiring the integrand be exact for 1 and for x. For one we have $\int_0^1 1\,dx = W_0 \cdot 1$ or $1 = W_0$. Moreover we require the integral to be exact for x because

$$\int_0^1 x\,dx = W_1 x_0 \cong \left.\frac{x^2}{2}\right|_0^1 = W_1 x_0.$$

$$\frac{1}{2} = W_1 x_0.$$

Since $W_1 = 1$ we have $x_0 = \frac{1}{2}$. Thus we have established the midpoint rule

$$\int_0^1 F(x)dx \cong F(1/2).$$

We next derive a quadrature formula of the following type

$$\int_{-1}^1 F(x)dx \cong W_1 F(x_1) + W_2 F(x_2)$$

Requiring the formula to be exact for $F(x) = 1$ we have

$$\int_{-1}^1 1\,dx = W_1 + W_2 \text{ or } W_1 + W_2 = 2.$$

Requiring the formula to be exact for x yields

$$\int_{-1}^1 x\,dx = \left.\frac{x^2}{2}\right|_{-1}^1 = 0$$

$$0 = W_1 x_1 + W_2 x_2.$$

For x^2 being exact we have

$$\int_{-1}^1 x^2\,dx = \left.\frac{x^3}{3}\right|_{-1}^1 = \frac{2}{3} = W_1 x_1^2 + W_2 x_2^2.$$

For x^3 being exact

$$\int_{-1}^1 x^3\,dx = 0 = W_1 x_1^3 + W_2 x_2^3$$

We thus have a system of four equations in the four unknowns W_1, W_2, x_1, x_2. We try for a solution with $W_1 = W_2$. Then $W_1 = W_2 = 1$, $x_1 = -x_2$ and $\frac{2}{3} = 2x_1^2$ or $x_1 = \frac{1}{\sqrt{3}}$, $x_2 = -\frac{1}{\sqrt{3}}$. The equations $0 = W_1 x_1^3 + W_2 x_2^3$ and $\frac{2}{3} = x_1^2 + x_2^2$ are thus also satisfied. Hence we have the rule

$$\int_{-1}^{1} F(x)dx \cong F\left(\frac{1}{\sqrt{3}}\right) + F\left(-\frac{1}{\sqrt{3}}\right)$$

exact for $1, x, x^2, x^3$.

Suppose we wished to find a formula for $\int_0^1 xF(x)dx \cong W_1 F(x_1)$. First we require it to be exact for $F(x) = 1$. We obtain

$$\int_0^1 x \cdot 1\, dx = \left.\frac{x^2}{2}\right|_0^1 = \frac{1}{2} = W_1 \cdot 1.$$

We then require it to be exact for $F(x) = x$

$$\int_0^1 x \cdot x\, dx = \left.\frac{x^3}{3}\right|_0^1 = \frac{1}{3} = W_1 x_1.$$

Hence

$$\frac{1}{3} = \frac{1}{2} x_1 \text{ or } x_1 = \frac{2}{3}.$$

Hence we have derived the integration rule

$$\int_0^1 x\, F(x)dx \cong \frac{1}{2} \cdot F\left(\frac{2}{3}\right).$$

Romberg Integration

An extension of trapezoidal rule which is sometimes used on a digital computer to compute the values of definite integrals is that of Romberg integration. A derivation of this algorithm would take us too far afield. We merely state the algorithm and give numerical examples.

Consider the trapezoidal rule for computation of $\int_a^b f(x)dx$. If $h = \frac{b-a}{2^k}$, then the trapezoidal rule yields

$$\frac{b-a}{2^k}\left[\frac{1}{2}f_0 + f_1 + f_2 + \cdots + f_{2^k-1} + \frac{f_{2^k}}{2}\right]$$ we denote this by $T_{0,k}$. Now if in general we denote by

$$T_{m,k} = \frac{1}{4^m - 1}\left(4^m T_{m-1,k+1} - T_{m-1,k}\right)$$

$k = 0, 1, \ldots$

$m = 1, 2, \ldots$

we have a table of the T's made up as follows:

$$\begin{array}{lll} T_{0,0} & & \\ T_{0,1} & T_{1,0} & \\ T_{0,2} & T_{1,1} & T_{2,0}\cdots \\ \vdots & & \\ T_{0,m} & T_{1,m-1}, \ldots T_{m,0} & \end{array}$$

We state that under certain conditions on f(x), $T_{m,0}$ is a better result than the trapezoidal rule.

Example

Compute $\int_0^1 e^x\,dx$ using Romberg integration.

$$T_{0,0} = \frac{1-0}{2^0}\left[\frac{e^0}{2} + \frac{e^1}{2}\right] = \frac{1 + 2.7183}{2}$$

$$= 1.85915 = 1.8592 \text{ to 4D}$$

$$T_{0,1} = \frac{1-0}{2}\left[\frac{1}{2}e^0 + e^{.5} + \frac{e^{1.}}{2}\right]$$

$$= \frac{1}{2}\left[.5 + 1.6487 + \frac{2.7183}{2}\right]$$

$$= 1.753925 = 1.7539 \text{ to 4D}$$

$$T_{0,2} = \frac{1-0}{4}\left[\frac{e^0}{2} + e^{.25} + e^{.5} + e^{.75} + \frac{e^{1.}}{2}\right]$$
$$= \frac{1}{4}[.5 + 1.2840 + 1.6487 + 2.1170 + 1.3592]$$
$$= 1.7272$$
$$T_{1,0} = \frac{1}{3}(4T_{0,1} - T_{0,0})$$
$$= \frac{1}{3}(4.(1.7539) - 1.8592)$$
$$= 1.7188$$
$$T_{1,1} = \frac{1}{3}(4T_{0,2} - T_{0,1})$$
$$= \frac{1}{3}(4.(1.7272) - 1.7639)$$
$$= 1.7183$$
$$T_{2,0} = \frac{1}{4^2-1}(4^2\, T_{1,1} - T_{1,0})$$
$$= \frac{16\cdot(1.7183) - 1.7188}{15}$$
$$= 1.7183 \text{ to 4D}$$

whereas

$$\int_0^1 e^x\, dx = e^1 - e^0 = 1.7183 \text{ to 4D}$$

Exercises

1. Write a fortran program using Romberg's method to calculate $\int_0^1 e^x\, dx$. (Calculate $T_{10,0}$.)
2. Write out Romberg's scheme for $T_{1,k}$. Compare this with Simpson's rule.
3. Compute $\int_0^1 \sin x\, dx$ using Romberg integration. In particular, compute $(T_{3,0})$.

Chapter 6

DIFFERENCES AND DIFFERENCE OPERATORS

The forward difference operator $\triangle$ is defined as $\triangle f(x) = f(x + h) - f(x)$.
For example $\triangle x^2 = (x + h)^2 - x^2 = 2xh + h^2$. $\triangle^2$ is defiued as

$$
\begin{aligned}
\triangle(\triangle f(x)) &= \triangle[f(x + h) - f(x)] \\
&= \triangle f(x + h) - \triangle f(x) \\
&= f(x + 2h) - f(x + h) - [f(x + h) - f(x)] \\
&= f(x + 2h) - 2f(x + h) + f(x).
\end{aligned}
$$

$\triangle^3 f(x)$ is defined as

$$
\begin{aligned}
\triangle^3 f(x) &= \triangle(\triangle^2 f(x)) \\
&= \triangle(f(x + 2h) - 2f(x + h) + f(x)) \\
&= \triangle f(x + 2h) - 2\triangle f(x + h) + \triangle f(x) \\
&= f(x + 3h) - f(x + 2h) - 2f(x + 2h) \\
&\qquad - f(x + h) + f(x + h) - f(x) \\
&= f(x + 3h) - 3f(x + h) + 3f(x + h) - f(x).
\end{aligned}
$$

$\triangle^n f(x)$ is defined inductively as $\triangle^n f(x) = \triangle(\triangle^{n-1} f(x))$.

A table of differences can be made up in the following manner. Suppose $f(x)$ is given at an equally spaced set of points $x_0, x_1, x_2, \cdots, x_n$, $x_i = x_0 + ih$, $i = 0, 1, 2, \cdots, n$. Then we have

Table 1

x	$f(x)$				
x_0	$f(x_0)$				
		$\triangle f(x_0)$			
x_1	$f(x_1)$		$\triangle^2 f(x_0)$		
		$\triangle f(x_1)$		$\triangle^3 f(x_0)$	
x_2	$f(x_2)$		$\triangle^2 f(x_1)$		$\triangle^4 f(x_0)$
		$\triangle f(x_2)$		$\triangle^3 f(x_1)$	
x_3	$f(x_3)$		$\triangle^2 f(x_2)$		
		$\triangle f(x_3)$			
x_4	$f(x_4)$				

Example

A table of $f(x) = e^x$ with $x_0 = 0, x_1 = .1, x_2 = .2, x_3 = .3, x_4 = .4$.

Table 2

x	e^x	$\triangle f(x)$	$\triangle^2 f(x)$	$\triangle^3 f(x)$	$\triangle^4 f(x)$
0	1.0000				
		.1052			
.1	1.1052		.0110		
		.1162		.0013	
.2	1.2214		.0123		− .0002
		.1285		.0011	
.3	1.3499		.0134		
		.1419			
.4	1.4918				

Difference tables are useful in conjuction with interpolation, numerical integration and numerical differentiation. In order to discuss interpolation we introduce the operator E defined as $E\,f(x) = f(x + h)$. $E^2 f(x)$ is defined as $E(E\,f(x)) = E\,f(x + h) = f(x + 2h)$. In general for an arbitrary real number α we define $E^\alpha f(x) = f(x + \alpha h)$.

Since $\triangle f(x) = f(x + h) - f(x)$ we have $\triangle f(x) = E\,f(x) - 1 \cdot f(x)$ where 1 is the identity operator, i.e., $1\,f(x) = f(x)$. Thus we obtain $E\,f(x) = \triangle f(x) + 1 \cdot f(x)$, which yields the identity $1 + \triangle = E$.

If we raise both sides of the identity $E = 1 + \triangle$ to the s power we obtain $E^s f(x) = (1 + \triangle)^s\, f(x)$ or $f(x + sh) = (1 + \triangle)^s f(x)$. Expanding $(1 + \triangle)^s$ in a Taylor's series we obtain

$$f(x + sh) = f(x) + s\triangle f(x) + \frac{s(s-1)}{2} \triangle^2 f(x)$$
$$+ \cdots + \frac{s(s-1)(s-2)\cdots(s-n+1)}{n!} \triangle^n f(x) + \text{Error} \qquad (1)$$

It can be shown that

$$\text{Error} = \frac{h^{n+1}}{(n+1)!} s(s-1)(s-2)\cdots(s-n)\, f^{n+1}(\xi).$$

Here $\xi \in [x_0, x_n]$, $x \in [x_0, x_n]$.

Example 1.

Find $f(.05) = e^{.05}$, using the above interpolation formula where the difference Table 2 is used.

Here $x_0 = 0$, $h = .1$, so $s = (x - x_0)/h = (.05 - 0)/.1 = .5$

Hence

$$f(.05) = f_0 + s\triangle f_0 + \frac{s(s-1)}{2}\triangle^2 f_0 + \frac{s(s-1)(s-2)}{3!}\triangle^3 f_0$$

$$+ \frac{s(s-1)(s-2)(s-3)}{4!}\triangle^4 f_0 + \text{Error}$$

$$f(.05) \approx 1 + \frac{1}{2}\cdot .1052 + \frac{1}{2}\cdot -\frac{1}{2}\cdot\frac{1}{2}(.0110) + \frac{1}{2}\left(-\frac{1}{2}\right)\left(-\frac{3}{2}\right)$$

$$\cdot\frac{1}{6}(.0013)$$

$$+\frac{1}{2}\cdot\frac{-1}{2}\cdot\frac{-3}{2}\cdot\frac{-5}{2}\cdot\frac{1}{4!}(-.0002)$$

$f(.05) \approx 1.051314 \approx 1.0513.$

The exact value to 4D is 1.0513.

Example 2.

Let $f(x) = x^2$, $x_0 = 0$, $x_i = x_0 + ih$, $h = 1$, $i = 0, 1, 2, 3, 4$. Then the difference Table 3 reads:

Table 3

x	f(x)	$\triangle f(x)$	$\triangle^2 f(x)$	$\triangle^3 f(x)$	$\triangle^4 f(x)$
0	0				
		1			
1	1		2		
		3		0	
2	4		2		0
		5		0	
3	9		2		
		7			
4	16				

Suppose we wish to calculate f(.75) by the interpolation formula. Here $x = x_0 + sh$, $x_0 = 0$, $h = 1$, $s = (.75 - 0)/1 = .75$. Thus

$$f(.75) \approx 0 + \frac{3}{4} \triangle\, f(x_0) + \frac{\left(\frac{3}{4}\right)\left(\frac{3}{4} - 1\right)}{2} \triangle^2 f(x_0) + 0$$

Since $\triangle^3 f(x_0) = 0$, $\triangle^4 f(x_0) = 0$, etc.,

$$f(.75) \approx \frac{3}{4} \cdot 1 + \frac{3}{4} \cdot - \frac{1}{4} \cdot \frac{1}{2} \cdot 2 \approx \frac{9}{16}$$

whereas in this case

$$\text{Error} = \frac{s(s-1)(s-2)(s-3)(s-4)}{51} f^v(\xi)$$

where $s = \frac{3}{4}$, but $f^v(\xi) = 0$. Hence the interpolated value is the exact value.

The backward difference operator ∇ is defined as $\nabla f(x) = f(x) - f(x - h)$. $\nabla^2 f(x)$ is defined as $\nabla^2 f(x) = \nabla(\nabla f(x))$
$= \nabla\,(f(x) - f(x - h))$
$= \nabla f(x) - \nabla f(x - h) = f(x) - f(x - h) - (f(x - h) - f(x - 2h))$
$= f(x) - 2f(x - h) + f(x - 2h)$.

In general $\nabla^n f(x)$ is defined as $\nabla^n f(x) = \nabla(\nabla^{n-1} f(x))$. In a table of differences ∇ is interpreted as follows.

Table 4

x	f(x)		
.			
.			
.			
x_{n-2}	$f(x_{n-2})$		
		$\nabla f(x_{n-1})$	
x_{n-1}	$f(x_{n-1})$		$\nabla^2 f(x_n)$
		$\nabla f(x_n)$ ↗	
x_n	$f(x_n)$ ↗		

From Table 4 we read $\nabla f(4) = 7$, $\nabla^2 f(4) = 2$, $\nabla^3 f(4) = 0$, $\nabla^4 f(4) = 0$.

An interpolating polynomial using backward differences can be constructed similar to the one for forward differences. We do not derive the result but merely state it here. Let $s = (x - x_n)/h$; then

$$f(x_n + sh) = f(x_n) + s\,\nabla f(x_n) + \frac{s(s+1)}{2}\nabla^2 f(x_n) + \cdots + \frac{s(s+1)\cdots(s+n-1)}{n!}\nabla^n f(x_n) + \text{Error} \quad (2)$$

where the

$$\text{Error} = \frac{h^{n+1}}{(n+1)!}\, s(s+1)(s+2)\cdots(s+n) f^{n+1}(\xi)$$

Here $\xi \,\epsilon [x_0, x_n]$.

Example

Calculate $f(.35) = e^{.35}$, using formula 2 and table 2. Here $x_n = .4$, $s = (.35 - .4)/.1 = -.5$. Thus

$$f(.35) \approx 1.4918 + -\frac{1}{2}\cdot(.1419) + -\frac{1}{2}\cdot\frac{1}{2}\cdot\frac{1}{2}\cdot(.0134)$$
$$+ -\frac{1}{2}\cdot\frac{1}{2}\cdot\frac{3}{2}\cdot\frac{1}{3!}.0011 + -\frac{1}{2}\cdot\frac{1}{2}\cdot\frac{3}{2}\cdot\frac{5}{2}\cdot\frac{1}{4!}(-.0002)$$
$$\approx 1.4191085 \approx 1.4191$$

The exact value to 4D is $1.4191 = e^{.35}$.

Example

Use Table 3 to calculate f(3.3). Here $x_n = 4$, $s = (3.3 - 4)/1 = -.7$. So formula 2 reads

$$f(3.3) \approx 16 + (-.7)\,7 + \frac{(-.7)(.3)}{2}\cdot 2 + 0$$
$$\approx 16 - 4.9 - .21 = 10.89.$$

Since the Error = 0 we have the exact result.

We shall next derive a formula for numerical differentiation. Expanding $f(x + h)$ in a Taylor's series about x we obtain

$$f(x+h) = f(x) + h\,f'(x) + \frac{h^2}{2}f''(x) + \frac{h^3}{3}f'''(x) + \cdots + \frac{h^n}{n!}f^n(x) + \cdots$$

or

$$f(x+h) = f(x) + h\frac{d}{dx}f(x) + \frac{h^2}{2}\frac{d^2}{dx^2}f(x) + \cdots + \frac{h^n}{n!}\frac{d^n}{dx^n}f(x) + \cdots$$

If we write this as

$$f(x+h) = \left(1 + h\frac{d}{dx} + \frac{h^2}{2}\frac{d^2}{dx^2} + \cdots + \frac{h^n}{n!}\frac{d^n}{dx^n} + \cdots\right) f(x)$$

we have

$$F\,f(x) = e^{h\frac{d}{dx}} f(x),$$

or the identity

$$E = e^{h\frac{d}{dx}}.$$

Since $1 + \Delta = E$, we have $1 + \Delta = e^{h\frac{d}{dx}}$. Taking logarithms of both sides we obtain

$$h\frac{d}{dx} = \log\,(1+\Delta).$$

Expanding $\log\,(1+\Delta)$ in a Taylor's series we have the identity

$$h\frac{d}{dx} = \left(\Delta - \frac{\Delta^2}{2} + \frac{\Delta^3}{3} - \frac{\Delta^4}{4} + \frac{\Delta^5}{5} - \cdots\right) \quad (3)$$

If we write this as

$$h\,f'(x_0) = \Delta f(x_0) - \frac{\Delta^2}{2}f(x_0) + \frac{\Delta^3}{3}f(x_0) - \frac{\Delta^4}{4}f(x_0) + \frac{\Delta^5 f(x_0)}{5} - \cdots$$

we can calculate an approximation to the derivative $f'(x_0)$. We note that if we retain only the first term in the series we obtain $h\,f'(x_0) \approx \Delta\,f(x_0)$ or

$$f'(x_0) \approx \frac{f(x_0+h) - f(x_0)}{h}.$$

Example

Use Table 2 to calculate an approximate value of $f'(0)$. Here $h = .1$, so we have

$$.1\, f'(0) \approx \left(.1052 - \frac{.0110}{2} + \frac{.0013}{3} - \frac{-.0002}{4} \right)$$

$$f'(0) \approx \frac{.0999875}{.1} \approx .9999$$

whereas the exact value is $f'(0) = e^0 = 1$.

We next derive numerical integration formula using differences. We define

$$J\, f(x) = \int_{x}^{x+h} f(t)\, dt.$$

Now it can be shown that $\frac{d}{dx} J\, f(x) = f(x+h) - f(x)$. Since $\triangle f(x) = f(x+h) - f(x)$ we have $\frac{d}{dx} J\, f(x) = \triangle f(x)$. Letting $\frac{d}{dx} = D$, we have the identity $D\, J = \triangle$.

Now

$$J\, f(x_0) = \int_{x_0}^{x_0+h} f(t)dt.$$

We can write

$$J\, f(x_0) = \frac{\triangle}{D} f(x_0) = \frac{h\triangle}{hD} f(x_0).$$

Since $h\, D = \log\,(1 + \triangle)$ we have

$$J\, f(x_0) = \frac{h \triangle f(x_0)}{\left(\triangle - \frac{\triangle^2}{2} + \frac{\triangle^3}{3} - \frac{\triangle^4}{4} + \frac{\triangle^5}{5} \cdots \right)}$$

Performing the division we obtain

$$J\, f(x_0) = \int_{x_0}^{x_0+h} f(t)dt$$

$$= h\left[1 + \frac{\triangle}{2} - \frac{\triangle^2}{12} + \frac{\triangle^3}{24} - \frac{19}{720} \triangle^4 + \cdots \right] f(x_0)$$

$$= h\left[f(x_0) + \frac{\triangle}{2} f(x_0) - \frac{\triangle^2}{12} f(x_0) + \frac{\triangle^3}{24} f(x_0) + \cdots \right] \quad (4)$$

If we retain just first differences in equation (4) we have

$$\int_{x_0}^{x_0+h} f(t)dt \approx h\left[f(x_0) + \frac{\Delta}{2} f(x_0) \right]$$

$$= \frac{h}{2}[2f(x_0) + f(x_0+h) - f(x_0)] = \frac{h}{2}[f(x_0) + f(x_0+h)]$$

This is just the trapezoidal rule.

Example

Integrate $\int_0^{.1} e^x\, dx$, using Table 2 and formula (4). We have

$$\int_0^{.1} e^x\, dx = .1\left[1 + \frac{1}{2}\cdot(.1052) - \frac{1}{12}(.0110) + \frac{1}{24}(.0013) - \frac{19}{720}(0.0002)\right]$$

$$= (1.05170).1 = .10517 \approx .1052$$

The exact value is $\int_0^{.1} e^x\, dx = e^{.1} - e^0 = .1052.$

In a manner similar to the one above, a numerical integration formula can be derived using backward differences. We do not derive the method here but merely state it.

$$\int_{x_n}^{x_n+h} f(t)dt = h\left(1 + \frac{1}{2}\nabla + \frac{5}{12}\nabla^2 + \frac{3}{8}\nabla^3 + \frac{251}{720}\nabla^4 + \frac{95}{288}\nabla^5 + \cdots\right) f(x_n) \qquad (5)$$

Evaluate $\int_{.4}^{.5} e^x\, dx$, using Table 2.

$$\int_{.4}^{.5} e^x\, dx \approx .1\left[1.4918 + \frac{1}{2}(.1419) + \frac{5}{12}(.0134) + \frac{3}{4}(.0011) + \frac{251}{720}(-.0002)\right]$$

$$= .1[1.4918 + .0795 + .005058 + .0004125 - .000069]$$

$$= .156815 \approx .1568$$

whereas the exact value to 4D is

$$\int_{.4}^{.5} e^x \, dx = e^{.5} - e^{.4} = 1.6487 - 1.4918 = .1569.$$

Exercises

1. Use Table 2, $f(x) = e^x$, and the forward difference interpolating polynomial to calculate $f(.09)$. Obtain a bound on the error.
2. Use Table 2, $f(x) = e^x$ and the forward difference interpolating polynomial to calculate $f(.15)$. (Hint: here $x_0 = .1$). Bound the error made.
3. Use Table 2, $f(x) = e^x$ and the backward difference interpolating polynomial to calculate $f(.32)$.
4. Use Table 2 and the numerical differentiation formula to calculate $f'(.1)$. (Hint: here $x_0 = .1$).
5. Use Table 3 and the numerical integration formula (4) to calculate $\int_0^1 x^2 \, dx$.
6. Use Table 3 and the numerical integration formula (5) to calculate $\int_4^5 x^2 \, dx$.

Section 6.2 The Euler-Maclaurin Sum Formula. In this section we shall derive an extension of the trapezoidal rule which may be applied to finite sums or integrals.

We begin by noting the relationship

$$\triangle f(x) = D\,J\,f(x) \tag{6.2.1}$$

From this we deduce that

$$h\,f(x) = \frac{hD}{e^{hD} - 1} J\,f(x) \tag{6.2.2}$$

Expanding e^{hD} we have

$$h\,f(x) = \frac{hD\,J\,f(x)}{hD + \frac{h^2D^2}{2} + \frac{h^3D^3}{6} + \cdots} \tag{6.2.3}$$

Simplifying we have

$$h\,f(x) = \frac{1}{1 - \frac{hD}{2} + \frac{h^2D^2}{6} + \cdots}\,J\,f(x) \tag{6.2.4}$$

Dividing through we have

$$h\,f(x) = \left(1 - \frac{hD}{2} + \frac{h^2D^2}{12} + \cdots\right) J\,f(x) \tag{6.2.5}$$

or

$$h\,f(x) = J\,f(x) - h\left[\frac{-f(x) + f(x+h)}{2}\right] + \frac{h^2}{12}[f'(x+h) - f'(x)] + \cdots \tag{6.2.6}$$

Letting $x_i = x_0 + ih$, $i = 0, 1, 2, \cdots r - 1$, we have

$$h\,f(x_0) = J\,f(x_0) - h\left[\frac{-f(x_0) + f(x_0+h)}{2}\right] + \frac{h^2}{12}\left[f'(x_0+h) - f'(x_0)\right] + \cdots$$

$$h\,f(x_1) = J\,f(x_1) - \frac{h}{2}[-f_1 + f_2] + \frac{h^2}{12}[f'_2 - f'_1] + \cdots$$

$$h\,f(x_{r-1}) = J\,f(x_{r-1}) - \frac{h}{2}[-f_{r-1} + f_r] + \frac{h^2}{12}(f'_r - f'_{r-1}] + \cdots$$

Adding up both sides we have

$$h \sum_{r=0}^{r-1} f(x_i) = \int_{x_0}^{x_r} f(x)dx + \frac{h}{2}[f_0 - f_r] + \frac{h^2}{12}[f'_r - f'_0] + \cdots \tag{6.2.7}$$

This is the Euler-Maclaurin sum formula. It relates a sum to the integral with end correction terms $\frac{h^2}{12}[f'_r - f'_0]$, etc. Written in another form we have

$$\int_{x_0}^{x_r} f(x)dx = h\left[\frac{f_0}{2} + f_1 + f_2 + \cdots + f_{r-1} + \frac{f_r}{2}\right] - \frac{h^2}{12}[f'_r - f'_0] + \cdots \tag{6.2.8}$$

which is just the trapezoidal rule plus end correction terms.

We apply the Euler-Maclaurin sum formula to find

$$\sum_{i=0}^{r-1} (i+1) = 1 + 2 + \cdots + r,$$

a problem encountered in college algebra. Here $f(x) = x + 1$, $f'(x) = 1$, $h = 1$.

Substituting in (6.2.7) we have

$$\sum_{i=1}^{r-1} (i+1) = \int_0^r (x+1)dx + \frac{1}{2}[1 - (r+1)]$$

$$+ \frac{1}{12}[1 - 1] + 0$$

or

$$\sum_{r=0}^{r-1} (i+1) = \frac{(x+1)^2}{2}\Bigg|_0^r - \frac{1}{2}r$$

$$= \frac{(r+1)^2}{2} - \frac{1}{2} - \frac{1}{2}r$$

$$= \frac{(r+1)^2}{2} - \frac{(r+1)}{2} = \frac{(r+1)}{2}(r+1-1)$$

$$= \frac{r(r+1)}{2}.$$

Exercises

1. By considering an addition term in the e^{hD} series, derive an additional term in formula (6.2.7).
2. Use the result of exercise 1 to find $\sum_{i=0}^{r-1} i^2 = ?$
3. Use formula (6.2.8) to calculate $\int_0^{.1} e^x\, dx$. (Here $h = .1$).

Chapter 7

THE NUMERICAL SOLUTION OF ORDINARY DIFFERENTIAL EQUATIONS

Section 7.1 Introduction. Ordinary differential equations occur in science and engineering. Many times these equations are too complicated to solve analytically so recourse must be made to numerical methods. In this chapter we shall study some of the more common methods used in the numerical solution of ordinary differential equations.

Section 7.2 Basic Concepts and Definition. By the solution of the ordinary differential equation $y' = f(x, y)$, $y(x_0) = y_0$, we mean a function $\phi(x)$ such that $\phi'(x) = f(x, \phi(x))$, $\phi(x_0) = y_0$. It is easy to see that if $f(x, y)$ is continuous on some interval $[x_0, b]$, then $\phi(x) \epsilon c^1 [x_0, b]$.

Examples

The differential equation $y' = y$, $y(0) = 1$ has a solution $\phi(x) = e^x$ as is easily verified. The differential equation $y' = 1 + y^2$, $y(0) = 0$ has the solution $\phi(x) = \tan(x)$. The differential equation $y' + ay = b$, $y(0) = c$ has the solution

$$\phi(x) = \left(c - \frac{b}{a} \right) e^{-ax} + \frac{b}{a}.$$

Section 7.3 Euler's Method. We first consider Euler's method. We wish a numerical solution at a discrete set of points x_0, $x_0 + h$, $x_0 + 2h, \cdots, x_0 + nh = b$; here $h > 0$. If we expand $y(x_{i+1})$ in a Taylor's series about x' we obtain

$$y(x_{i+1}) = y(x_i) + hy'(x_i) + \frac{h^2}{2} y''(\xi) \tag{7.3.1}$$

Letting the approximate value of $x_0 + ih$ be represented by y_i we have, neglecting the second derivative term,

$$y_{i+1} = y_i + h\, f(x_i, y_i) \tag{7.3.2}$$

We thus have the algorithm.

Section 7.3 Algorithm. Let x_0, y_0, b, h, N, be given. Generate y_i according to the scheme

$$y_{i+1} = y_i + hf(x_i, y_i)$$
$$x_{i+1} = x_i + h$$

Example 1.

$$y' = y,\ y(0) = 1,\ h = .1,\ N = 3$$
$$y_1 = y_0 + hy_0 = 1.1$$
$$x_1 = .10$$
$$y_2 = y_1 + hy_1 = 1.1 + .1(1.1) = 1.21$$
$$x_2 = .2$$
$$y_3 = y_2 + hy_2 = 1.21 + .1(1.21) = 1.331.$$
$$x_3 = .3$$

while the exact values are

$$y(x_1) = 1.1052,\ y(x_2) = 1.2214,\ y(x_3) = 1.3499$$

Example 2.

$$y' = 1 + y^2,\ y(0) = 0,\ h = .1,\ N = 2$$

$$y_1 = y_0 + h(1 + y_0^2) = 0 + .1(1 + 0) = .1$$
$$x_1 = .1$$

$$y_2 = y_1 + h(1 + y_1^2) = .1 + .1(1 + .1) = .21$$
$$x_2 = .2$$

The exact values are $y(.1) = .10033$, $y(.2) = .20271$.

Euler's method is quite easily programmed. Consider the differential equation $y' = 1 + y^2 + x^2, y(0) = 0, h = .02, N = 50.$ The program is:

```
F(X, Y) = 1. + Y * Y + X * X
X = 0.
Y = 0.
```

```
      WRITE (6, 10)X, Y
      H = .02
      DO 1 I = 1, 50
      Y = Y +H * F(X, Y)
      X = X + H
      WRITE (6, 10) X, Y
   10 FORMAT (F10.2, E20.8)
    1 CONTINUE
      STOP
      END
```

Exercises

1. Solve $y' = 2y$, $y(0) = 1$, exact solution $y(x) = e^{2x}$, with $h = .1$, $N = 3$ by Euler's method. Compare the exact solution.
2. Solve by Euler's method $y' = 1 + y^2$, $y(0) = 0$, with $h = .05$, $N = 3$. Compare wiih the exact solution $y(x) = \tan(x)$.
3. Solve $y' + y = 2$, $y(0) = 2$ with $h = .05$, $N = 3$ by Euler's method. Compare with the exact solution.
4. Write a Fortran program to compute the numerical solution of $y' = 1 + y^2$, $y(0) = 0$, $h = .02$, $N = 25$, using Euler's method.

Section 7.4 Heun's Method. Euler's method is not accurate enough for many practical problems. Therefore we turn to a more accurate method. The derivation proceeds as follows.

If we integrate the equation $y' = f(x, y)$, $y(x_0) = y_0$ from x_i to x_{i+1} we obtain

$$y(x_{i+1}) = y(x_i) + \int_{x_i}^{x_{i+1}} f(x, y)dx. \tag{7.4.1}$$

We now approximate the integral in (7.4.1) by the trapezoidal rule to obtain

$$y(x_{i+1}) = y(x_i) + \frac{h}{2}[f(x_i, y(x_i)) + f(x_{i+1}, y(x_{i+1}))] + O(h^3). \tag{7.4.2}$$

We run into a new difficulty here, $y(x_{i+1})$, the unknown function occurs in the right-hand side of the equation. To avoid this difficulty we approximate $y(x_{i+1})$ by $y_i + h\, f(x_i, y_i)$ (Euler's method). Thus we obtain, by dropping the $O(h^3)$ term and substituting y_i for $y(x_i)$, the following

$$y_{i+1} = y_i + \frac{h}{2}[f(x_i, y_i) + f(x_{i+1}, y_i + h\, f(x_i, y_i))] \qquad (7.4.3)$$

This is Heun's method. We have the algorithm.

Section 7.4 Algorithm. Let x_0, y_0, h, N be given. To calculate y_i form the expresssions

$$y_{i+1} = y_i + \frac{h}{2}[f(x_i, y_i) + f(x_{i+1}, y_i + h\, f(x_i, y_i))]$$

$$x_{i+1} = x_i + h.$$

Example 1.

Calculate the solution of $y' = y$, $y(0) = 1$, $h = .1$, $N = 2$, by Heun's method.

We have

$$y_1 = y_0 + \frac{h}{2}[f(x_0, y_0) + f(x_1, y_0 + hy_0)]$$

$$y_1 = 1 + \frac{.1}{2}[y_0 + y_0 + hy_0]$$

$$= 1 + .1(y_0) + .005y_0$$

$$= 1.1050$$

$$x_1 = .1$$

$$y_2 = y_1 + \frac{h}{2}[f(x_1, y_1) + f(x_2, y_1 + hy_1)]$$

$$= 1.1050 + \frac{.1}{2}[1.1050 + 1.1050 + .1(1.1050)]$$

$$= 1.221025$$

$$x_2 = .2$$

Example 2.

Calculate the solution of $y' = 1 + y^2$, $y(0) = 0$, $h = .1$, $N = 2$, by Heun's method.

$$y_1 = y_0 + \frac{h}{2}[f(x_0, y_0) + f(x_0 + h, y_0 + hf(x_0, y_0))]$$

$$= 0 + \frac{.1}{2}[1 + f(.1, .1)]$$

$$= \frac{.1}{2}[1 + 1 + (.1)^2]$$

$$= \frac{.1}{2}[2.01] = .10050.$$

$$x_1 = .1$$

$$y_2 = y_1 + \frac{h}{2}[f(.1, .1005) + f(.1, .1005 + .11005)]$$

$$= .10050 + \frac{.1}{2}[1.01010025 + 1.044331302]$$

$$= .20322$$

$$x_2 = .2$$

A Fortran program to use Heun's method for the numerical solution of $y' = 1 + y^2$, $y(0) = 0$ follows. Here $h = .05$, $N = 20$.

```
      F(X, Y) = 1. + Y ** 2
      X = 0
      Y = 0
      WRITE (6, 10) X, Y
      H = .05
      N = 20
      DO 1 I = 1, N
      Z = F(X, Y)
      Y1 = Y + H * Z
      Y = Y + H/2, *(Z + F(X + H, Y1))
      X = X + H
      WRITE (6, 10) X, Y
   10 FORMAT (F10.2, E20.8)
    1 CONTINUE
      STOP
      END
```

Exercises

1. Solve, using Heun's method, $y' = 2y$, $y(0) = 1$, $h = .1$, $N = 3$. Compare with the exact solution, $y(x) = e^{2x}$.
2. Solve, using Heun's method, $y' = 1 + y^2$, $y(0) = 0$, $h = .05$, $N = 2$. Compare with the exact solution, $y(x) = \tan(x)$.
3. Write a Fortran program to calculate y, using Heun's method, when $y' = x^2 + y^2$, $y(0) = 1$, $h = .05$, $N = 20$.

Section 7.5 Taylor's Series. In the two previous methods Euler's method has truncation error $O(h^2)$, Heun's method $O(h^3)$. In this section we shall give a method which can at least theoretically give arbitrarily high order of h.

We illustrate the Taylor's series approach with a numerical example, $y' = y$, $y(0) = 1$, $h = .1$, $N = 2$. Differentiating, we obtain $y'' = y' = y$, $y''' = y'' = y$, $y^{iv} = y''' = y$, $y^{v} = y^{iv} = y$. Thus we have the Taylor expansion

$$y(h) = y(0) + hy'(0) + \frac{h^2}{2}y''(0) + \frac{h^3}{6}y'''(0) + \frac{h^4}{24}y^{iv}(0)$$
$$+ \frac{h^5}{120}y^{v}(0) + \frac{h^6}{720}y^{vi}(\xi). \quad 0 < \xi < h$$

Neglecting the truncation error term, we have the approximate value, after substituting in $y' = y$, $y'' = y$, $y''' = y$, etc.

$$y_1 = y_0 + hy_0 + \frac{h^2}{2}y_0 + \frac{h^3}{6}y_0 + \frac{h^4}{24}y_0 + \frac{h^5}{120}y_0$$

or

$$y_1 = 1 + .1 + \frac{(.1)^2}{2} + \frac{(.1)^3}{6} + \frac{(.1)^4}{24} + \frac{(.1)^5}{120}$$

$$y_1 = 1.10517088$$

For $\quad x_2 = x_0 + 2h$

$$y_2 = y_1 + hy_1 + \frac{h^2}{2}y_1 + \frac{h^3}{6}y_1 + \frac{h^4}{24}y_1 + \frac{h^5}{120}y_1$$
$$= \left(1 + h + \frac{h^2}{2} + \frac{h^3}{6} + \frac{h^4}{24} + \frac{h^5}{120}\right)y_1$$
$$= (1.10517088)\,(1.10517088)$$
$$= 1.22141442295$$

Example 2.

Apply Taylor's series through the third derivative to $y' = y^2 + x^2$, $y(0) = 1$, $h = .1$, $N = 1$.

$$y'' = 2yy' + 2x$$
$$y''' = 2(y')^2 + 2yy'' + 2$$

Then

$$y_1 = y_0 + hy'_0 + \frac{h^2}{2} y''_0 + \frac{h^3}{3} y'''_0$$

Now

$$y'_0 = 1,\ y''_0 = 2,\ y'''_0 = 8.$$

So

$$y_1 = 1 + .1 + \frac{(.1)^2}{2} \cdot 2 + \frac{(.1)^3}{6} \cdot 8$$
$$y_1 = 1.11133$$
$$x_1 = .1$$

Example 3.

Apply Taylor's series to the differential equation $y'' = -y$, $y(0) = 0$, $y'(0) = 1$, $h = .1$, $N = 2$, through the first five terms. Then

$$y''' = -y',\ y^{iv} = -y'' = y,\ y^{v} = -y''' = y'.$$
$$y_1 = y_0 + hy'_0 + \frac{h^2}{2} y''_0 + \frac{h^3}{6} y'''_0 + \frac{h^4}{24} y^{iv}_0 + \frac{h^5}{120} y^{v}_0$$
$$y_1 = y_0 - \frac{h^2}{2} y_0 + \frac{h^4}{24} y_0 + h\,y'_0 - \frac{h^3}{6} y'_0 + \frac{h^5}{120} y'_0$$
$$y_1 = h\,y'_0 - \frac{h^3}{6} y'_0 + \frac{h^5}{120} y'_0 = .1 - \frac{(.1)^3}{6} + \frac{(.1)^5}{120}$$
$$y_1 = .1 - .000166666 + .000000083333$$
$$= .09983342$$
$$x_1 = .1$$

Now we come to a new development. In order to calculate the solution at x_2, we must know y'_1. We calculate it using the Taylor's series.

$$y'_1 = y'_0 + hy''_0 + \frac{h^2}{2} y'''_0 + \frac{h_3}{6} y^{iv}_0 + \frac{h^4}{24} y^{v}_0$$

$$y'_1 = 1 - \frac{(.1)^2}{2} + \frac{(.1)^4}{24} 1$$

$$y'_1 = 1 - .005 + .00000416 = .99500416$$

$$y_2 = y_1 - \frac{h^2}{2} y_1 + \frac{h^4}{24} y_1 + hy'_0 - \frac{h^3}{6} y'_0 + \frac{h^5}{120} y'_0$$

$$= \left(1 - \frac{h^2}{2} + \frac{h^4}{24} \right) y_1 + \left(h - \frac{h^3}{6} + \frac{h^5}{120} \right) y'_1$$

$$= (.99500416)\,(.09983342) + (.09983342)(.99500416)$$

$$= 2(.99500416)\,(.09983342) = .198559$$

Use Example 2 to write a Fortran program using Taylor's series. In Example 2 we have $y' = y^2 + x^2, y(0) = 1, h = .1$. Then

$$y'' = 2yy' + 2x$$

$$y''' = 2(y')^2 + 2yy'' + 2$$

The Fortran program is H = .1, N = 10.

```
      N = 10
      H = .1
      X = 0.
      Y = 1.
      WRITE (6, 10) X, Y
      DO 1 I = 1, 10
      D 1 Y = X * X + Y * Y
      D 2 Y = 2. * Y * D 1 Y + 2 * X
      D 3 Y = 2. * D 1 Y ** 2 + 2. * X * D 2 Y + 2.
      Y = Y + H * D 1 Y + H * H/2. * D 2 Y + H ** 3/6. * D 3 Y
      X = X + H
      WRITE (6, 10) X, Y
10    FORMAT (F10.2, E20.8)
 1    CONTINUE
      STOP
      END
```

Exercises

1. Use Taylor's series through the first four terms to calculate the solution of $y' = y + x$, $h = .1$, $y(0) = 1$, $N = 2$.
2. Use Taylor's series through the first five terms to calculate the solution of $y' = y$, $y(0) = 1$, $h = .05$, $N = 3$.
3. Use Taylor's series through the first three terms to calculate the solution of $y' = 1 + y^2$, $y(0) = 0$, $h = .1$, $N = 2$.
4. Write and run a Fortran program to calculate the solution of $y' = 1 + y^2$, $y(0) = 0$, $h = .05$. Take the first four terms of the Taylor's series as your approximation $N = 10$. Compare with the exact solution $y(x) = \tan(x)$.
5. Write and run a Fortran program to calculate the solution of $y' = x^2 + y^4$, $y(0) = 0$, $h = .05$, $N = 10$. Use the first four terms of the Taylor's series.

Section 7.6 Method of Runge-Kutta. Taking derivatives for Taylor's series may be rather tedious and subject to mechanical errors. For this reason Runge devised a method which involves successive substitutions in f. The method agrees with the Taylor's series of y through the first four terms. We state the algorithm here.

Section 7.6 Algorithm. For $y' = f(x, y)$, $y(x_0) = y_0$, calculate $y_1, y_2, \cdots y_n$ according to

$$k_1 = f(x_i, y_i)$$

$$k_2 = f\left(x_i + \frac{h}{2}, y_i + \frac{h}{2} k_1\right)$$

$$k_3 = f\left(x_i + \frac{h}{2}, y_i + \frac{h}{2} k_2\right)$$

$$k_4 = f(x_i + h, y_i + hk_3)$$

$$y_{i+1} = y_i + \frac{h}{6} [k_1 + 2k_2 + 2k_3 + k_4]$$

$$x_{i+1} = x_i + h.$$

Example

Apply Runge-Kutta's method to $y' = y$, $y(0) = 1$, $h = .1$, $N = 1$.

$$k_1 = f(x_0, y_0) = y_0 = 1$$

$$k_2 = f\left(x_0 + \frac{h}{2}, y_0 + \frac{h}{2} k_1\right) = y_0 + \frac{h}{2} k_1 = 1.05$$

$$k_3 = f\left(x_0 + \frac{h}{2}, y_0 + \frac{h}{2} k_2\right) = y_0 + \frac{h}{2} k_2 = 1.0525$$

$$k_4 = f(x_0 + h, y_0 + hk_3) = y_0 + hk_3 = 1 + .1(1.0525) = 1.10525$$

$$y_1 = y_0 + \frac{h}{6}[k_1 + 2k_2 + 2k_3 + k_4]$$

$$= 1 + \frac{.1}{6}[1 + 1.05 \cdot 2 + 1.0525 \cdot 2 + 1.10525]$$

$$= 1.1051708$$

$$x_1 = x_0 + h = .1$$

A Fortran program for the Runge-Kutta method is as follows. Consider $y' = y^2 + \sin x$, $y(0) = 0$, $h = .05$, $N = 20$.

```
      REAL K1, K2, K3, K4
      F(X, Y) = Y * Y + SIN(X)
      X = 0.
      H = .05
      Y = 0.
      WRITE (6, 10)X, Y
      N = 20
      DO 1 I = 1, N
      K1 = F(X, Y)
      K2 = F(X + H/2., Y + H/2 * K1)
      K3 = F(X + H/2, Y + H/2. * K2)
      K4 = F(X + H, Y + K3)
      Y = Y + H/6. * (K1 + 2. * (K2 + K3) + K4)
      X = X + H
      WRITE (6, 10) X, Y
   10 FORMAT (F10.2, E20.8)
    1 CONTINUE
      STOP
      END
```

Exercises

1. Apply the Runge-Kutta method to $y' = y + x$, $h = .1$, $N = 2$, $y(0) = 1$.
2. Apply the Runge-Kutta method to $y' = 1 + y^2$, $y(0) = 0$, $N = 2$.
3. Apply the Runge-Kutta method to $y' = y$, $y(0) = 1$, $h = .05$, $N = 4$.
4. Write and run a Fortran program for $y' = y$, $y(0) = 1$, $h = .05$, $N = 20$, using Runge-Kutta's method. Print out your results and the exact results.
5. Write and run a Fortran program for $y' = x^2 + y^2$, $y(0) = 0$, $h = .05$, $N = 10$, using Runge-Kutta's method.

Section 7.7 The Numerical Solution of Systems of Ordinary Differential Equations. In this section we shall consider systems of differential equations of the form

$$\begin{aligned} y'_1 &= f_1(x, y_1, y_2, \cdots, y_n) \\ y'_2 &= f_2(x, y_1, y_2, \cdots, y_n) \\ y'_n &= f_n(x, y_1, y_2, \cdots, y_n) \end{aligned} \tag{1}$$

where $y_{10} = y_1(x_0)$, $y_{20} = y_2(x_0)$, $y_{n0} = y_n(x_0)$.
An example of a system of ODE is

$$\begin{aligned} y'_1 &= y_2 \\ & \qquad\qquad y_1(0) = 0,\ y_2(0) = 1 \\ y'_2 &= -y_1 \end{aligned}$$

The solution is $y_1(x) = \sin(x)$, $y_2(x) = \cos(x)$. This is easily verified since

$$\begin{aligned} y'_1 &= \cos(x) = y_2 & \qquad y_1(0) &= \sin(0) = 0 \\ y'_2 &= -\sin x = -y_1 & \qquad y_2(0) &= \cos(0) = 1 \end{aligned}$$

Systems of ODE also arise from the following observation. Consider the nth order ODE $y^n = f(x, y, y', y'', \cdots, y^{n-1})$. Then let

$$\begin{aligned} y_1 &= y \\ y_2 &= y'_1 = y' \\ y_3 &= y'_2 = y'' \\ y_n(x) &= y'_{n-1} = y^{n-1} \end{aligned}$$

Then we have the equivalent set of n first-order equations for the nth order ODE,

$$
\begin{aligned}
y'_1 &= y_2 \\
y'_2 &= y_3 \\
&\cdot \\
&\cdot \\
&\cdot \\
y'_{n-1} &= y_n \\
y'_n &= f(x, y_1, y_2, \cdots, y_n)
\end{aligned}
$$

Example

Convert the differential equation $y'' = -y$ to a system of first-order equations. We have $y = y_1$, $y'_1 = y_2$. Thus

$$
\begin{aligned}
y'_1 &= y_2 \\
y'_2 &= -y_1.
\end{aligned}
$$

Example

Convert the system $y''' = f(x, y, y'_1, y'')$ to a system of first-order equations. We have

$$
\begin{aligned}
y'_1 &= y_2 \\
y'_2 &= y_3 \\
y'_3 &= f(x, y_1, y_2, y_3)
\end{aligned}
$$

Thus the third-order ODE becomes a system of three first-order ODE's.

Euler's method for the system (1) of ODE's is as follows :

$$
\begin{aligned}
y_{1i+1} &= y_{1i} + h f_1(x_1, y_{1i}, y_{2i}, \cdots, y_{ni}) \\
y_{2i+1} &= y_{2i} + h f_2(x_i, y_{1i}, y_{2i}, \cdots, y_{ni}) \\
&\cdot \\
&\cdot \\
&\cdot \\
y_{ni+1} &= y_{ni} + h f_n(x_i, y_{1i}, y_{2i}, \cdots, y_{ni})
\end{aligned}
$$

where y_{1i} is the approximate value of $y_1(x_i)$, etc.

As an example we compute approximate values for the system

$$
\begin{aligned}
y'_1 &= y_2 \\
& \qquad\qquad y_1(0) = 0,\ y_2(0) = 1 \\
y'_2 &= -y_1
\end{aligned}
$$

where h = .1, N = 2. Here

$$y_{11} = y_{10} + hy_{20} = 0 + .1(1) = .1$$
$$y_{21} = y_{20} + h(-y_{10}) = 1 + .1(0) = 1$$
$$x = .1$$
$$y_{12} = y_{11} + h\,y_{21} = .1 + .1(1) = .2$$
$$y_{22} = y_{21} + h(-y_{11}) = 1 + -.01 = .99$$
$$x = .2$$

A Fortran program using Euler's method for the above system with h = .01, N = 50 is as follows:

```
Program Euler
   F1(X, Y1, Y2) = Y2
   F2(X, Y1, Y2) = - Y1
   X = 0
   H = .01
   N = 50
   Y1 = 0.
   Y2 = 1.0
   WRITE (6, 60) X, Y1, Y2
   DO 1 I = 1, N
   Z1 = Y1 + H * F1(X, Y1, Y2)
   Z2 = Y2 + H * F2(X, Y1, Y2)
   X = X + H
   Y1 = Z1
   Y2 = Z2
   WRITE (6, 60) X, Y1, Y2
60 FORMAT (F10.2, E20.8)
 1 CONTINUE
   STOP
   END
```

The method of Taylor's series may also be applied systems of ODE. We give an example of the system:

$$y'_1 = y_2 \qquad\qquad y_1(0) = 0,\ y_2(0) = 1$$
$$y'_2 = -y_1$$

Then

$$y''_1 = y'_2$$
$$y''_2 = -y'_1$$

So we have

$$y_{11} = y_{10} + hy'_{10} + \frac{h^2}{2} y''_{10} = y_{10} + hy_{20} + \frac{h^2}{2} y'_{20}$$

$$y_{21} = y_{20} + hy'_{20} + \frac{h^2}{2} y''_{20} = y_{20} + h(-y_{10}) - \frac{h^2}{2} y'_{10}$$

For h = .1, N = 1, we have

$$y_{11} = y_{10} + hy_{20} + \frac{h^2}{2} y'_{20} = .1(1) + \frac{.01}{2}(0) = .1$$

$$y_{21} = 1 + .1(0) + -\frac{.01}{2}(1) = .995$$

A Fortran program for the system of ODE using the Taylor's series method, taking the first four derivatives, is as follows:

$$y'_1 = y_2 \qquad y'''_1 = y''_2$$

$$y'_2 = -y_1 \qquad y'''_2 = -y''_1$$

$$y''_1 = +y'_2 \qquad y^{iv}_1 = y'''_2$$

$$y''_2 = -y'_1 \qquad y^{iv}_2 = -y'''_1$$

Program Taylor

```
X = 0.
Y1 = 0.
Y2 = 1.
WRITE (6,60) X, Y1, Y2
H = .01
N = 100
DO 1 I = 1, N
D1Y1 = Y2
D1Y2 = - Y1
D2Y1 = D1Y2
D2Y2 = - D1Y1
D3Y1 = D2Y2
D3Y2 = - D2Y1
D4Y1 = D3Y2
D4Y2 = - D3Y1
Y1 = Y1 + H * D1Y1 + H * H/2, * D2Y1
   + H ** 3/6. * D3Y1 + H ** 4/24. * D4Y1
Y2 = Y2 + H * D1Y2 + H * H/2, * D2Y2
   + H ** 3/6. * D3Y2 + H ** 4/24. * D4Y2
X = X + H
```

```
      WRITE (6, 60) X, Y1, Y2
   60 FORMAT (F10.2, 2E20.8)
    1 CONTINUE
      STOP
      END
```

The method of Runge-Kutta can be programmed for systems of differential equations. We limit ourselves to a two-by-two system of equations here of the form

$$y'_1 = f_1(x, y_1, y_2)$$
$$y'_2 = f_2(x, y_1, y_2)$$
$$y_{10} = y_1(x_0),\ y_{20} = y_2(x_0).$$

We have the following:

$$k_{11} = f_1(x_i, y_{1i}, y_{2i})$$
$$k_{21} = f_2(x_i, y_{1i}, y_{2i})$$
$$k_{12} = f_1(x_i + \frac{h}{2}, y_{1i} + \frac{h}{2} k_{11}, y_{2i} + \frac{h}{2} k_{21})$$
$$k_{22} = f_2(x_i + \frac{h}{2}, y_{1i} + \frac{h}{2} k_{11}, y_{2i} + \frac{h}{2} k_{21})$$
$$k_{13} = f_1(x_i + \frac{h}{2}, y_{1i} + \frac{h}{2} k_{12}, y_{2i} + \frac{h}{2} k_{22})$$
$$k_{23} = f_2(x_i + \frac{h}{2}, y_{1i} + \frac{h}{2} k_{12}, y_{2i} + \frac{h}{2} k_{22})$$
$$k_{14} = f_1(x_i + h, y_{1i} + hk_{13}, y_{2i} + hk_{23})$$
$$k_{24} = f_2(x_i + h, y_{1i} + hk_{13}, y_{2i} + hk_{23})$$
$$y_{1i+1} = y_{1i} + \frac{h}{6}(k_{11} + 2k_{12} + 2k_{13} + k_{14})$$
$$y_{2i+1} = y_{2i} + \frac{h}{6}(k_{21} + 2k_{22} + 2k_{23} + k_{24})$$

A Fortran program to illustrate the method is as follows: consider $y'' = -y$, $y(0) = 0$, $y'(0) = 1$. Then $y'_1 = y_2$, $y'_2 = -y_1$, and $y_1(0) = 0$, $y_2(0) = 1$. $h = .01$, $N = 100$.

```
      REAL K11, K21, K12, K22, K13, K23, K14, K24
      F1(X, Y1, Y2) = Y2
      F2(X, Y1, Y2) = - Y1
      H = .01
      X = 0.
      Y1 = 0.
      Y2 = 1.
```

```
      WRITE (6, 60) X, Y1, Y2
      N = 100
      DO 1 I = 1, N
      K11 = F1(X, Y1, Y2)
      K21 = F2(X, Y1, Y2)
      K12 = F1(X + H/2., Y1 + H/2. * K11, Y2 + H/2. * K21)
      K22 = F2(X + H/2., Y1 + H/2. * K11, Y2 + H/2. * K21)
      K13 = F1(X + H/2., Y1 + H/2. * K12, Y2 + H/2. * K22)
      K23 = F2(X + H/2., Y1 + H/2. * K12, Y2 + H/2. * K22)
      K14 = F1(X + H, Y1 + H * K13, Y2 + H * K23)
      K24 = F2(X + H, Y1 + H * K13, Y2 + H * K23)
      Y1 = Y1 + H/6. * (K11 + 2. * (K12 + K13) + K14)
      Y2 = Y2 + H/6. * (K21 + 2. * (K22 + K23) + K24)
      X = X + H
      WRITE (6, 60) X, Y1, Y2
   60 FORMAT (F10.2, 2E20.8)
    1 CONTINUE
      STOP
      END
```

An example to show what might occur in the process of solving ODE is the following. Consider the differential equation $y'' = 100y$, $y(0) = 1$, $y'(0) = -10$. Clearly the solution is $y(x) = e^{-10x}$.

Taking Taylor's series through the first five derivatives and h as .02, we compute the solution and obtain

x	y	exact
.2	.1353342	.13533550
.4	.018314140	.018315700
.6	.0024694030	.0024787660
.88	$-.2 \times 10^{-5}$	$.1507 \cdot 10^{-3}$

Thus the numerical solution is negative at .88 but $y(x) > 0$; hence we are in error.

The reason for this is as follows: at each step we are solving an initial value problem. There will be some error in the solution. The general solution of $y'' = 100y$ is $u(x) = C_1 e^{10x} + C_2 e^{-10x}$. We are seeking the solution where $C_1 = 0$, $C_2 = 1$, but instead we get $C_1 = \pm \epsilon_1$, $C_2 = 1 \pm \epsilon_2$ and the solution e^{10x} grows quite rapidly and thus "swamps" the lower curve e^{-10x}. No numerical method for solving ODE's will work on such an example.

Exercises

1. Convert the ODE $y'' = 6x + 9y' + 10y$, $y(0) = 1$, $y'(0) = 0$, into a system of first-order equations.
2. Convert the system of ODE's $y''_1 = f_1(x, y_1, y_2)$, $y''_2 = f_2(x, y_1, y_2)$ into a system of four first-order ODE's.
3. Solve the system

   $y'_1 = y_1 + y_2$

   $y'_2 = y_2 - y_1$

   $y_1(0) = 0$, $y_2(0) = 1$, using Euler's method with $h = .1$, $n = 2$.
4. Write a Fortran program using the ODE of Exercise 3 with $h = .02$, $n = 100$, using Euler's method. Compare with the exact solution $y_1 = e^x \sin x$, $y_2 = e^x \cos x$.
5. Write a Fortran program using Taylor's series through the first four derivatives to calculate an approximate solution to the ODE system given in Exercise 3 with $h = .02$, $n = 100$. Compare with the exact solution $y_1 = e^x \sin x$, $y_2 = e^x \cos x$.
6. Write a Fortran program using the method of Runge-Kutta for the system given in Exercise 3 with $h = .02$, $n = 100$. Compare with the exact solution.

Section 7.8 Predictor-Connector Methods. Consider the differential $y' = f(x, y)$, $y(x_0) = y_0$. Integrate between x_r and x_{r+1} to obtain

$$y(x_{r+1}) = y(x_r) + \int_{x_r}^{x_{r+1}} f(x, y)dx.$$

Then use the trapezoidal rule in the resulting integral to obtain

$$y_{r+1} = y_r + \frac{h}{2}[f(x_r, y_r) + f(x_{r+1}, y_{r+1})]$$

We note a new difficulty has arisen. We want to solve for y_{r+1} but it is on the right-hand side entering in a usually non-linear fashion. To avoid this difficulty we introduce a predictor value $y^*_{r+1} = y_r + h\,f(x_r, y_r)$, using Euler's method. We then substitute into the corrector

$$y^{(i)}_{r+1} = y_r + \frac{h}{2}[f(x_r, y_r) + f(x_{r+1}, y^{i-1}_{r+1})]$$

to iterate for a new value. The following example illustrates the method :

Example

Solve the differential equation $y' = y$, $y(0) = 1$, using the predictor-corrector method with $h = .1$.

$$y^*_1 = y_0 + h\,y_0 = 1 + .1 \cdot 1 = 1.1$$

$$y^{(0)}_1 = 1.1$$

$$y^1_1 = y_0 + \frac{h}{2}[y_0 + y^{(0)}_1] = 1 + .05\,[1 + 1.1]$$

$$y^{(1)}_1 = 1.105$$

$$y^{(2)}_1 = 1. + .05\,[1 + 1.105] = 1.10525$$

We note the above procedure gives us no indication as to when to stop iterating. We now turn to a method which will give us such an indication. We use the predictor-corrector formula

$$y^*_{r+1} = y_r + \frac{h}{2}[3f(x_r, y_r) - f(x_{r-1}, y_{r-1})]$$

$$y^{(j)}_{r+1} = y_r + \frac{h}{2}[f(x_r, y_r) + f(x_{r+1}, y^{j-1}_{r+1})]$$

Then solve

$$y(x_{r+1}) - y^*_{r+1} = \frac{5h^3}{12} f''_r + \cdots y(x_{r+1}) - y^{(j)}_{r+1}$$

$$= \frac{h^3}{12} f''_r + \cdots$$

We have

$$y^{j}_{r+1} - y^{*}_{r+1} = \frac{h^3}{2} f''_r + \cdots$$

or that

$$y^{j}_{r+1} - y^{*}_{r+1} \approx \frac{1}{6} [y(x_{r+1}) - y^{j}_{r+1}]$$

Thus we have a criterion to quit iterating. The error of $y^{j}_{r+1} - y^{*}_{r+1}$ is $\frac{1}{6}$ of $(y(x_{r+1}) - y^{j}_{r+1})$.

We note in starting the process for the algorithm under consideration a new difficulty. That is in solving for $y^{*}_2 = y_1 + \frac{h}{2} [3f(x_1, y_1) - f(x_0, y)$ that y_1 is not given. We must solve for y_1 using either the Taylor's series or Runge-Kutta type methods. Using Taylor's series we have

$$y_1 = y_0 + hy_0 + \frac{h^2}{2} y''_0.$$

We can then continue the solution forward. A Fortran program for the following example is given :

Example

Write a Fortran program to compute the solution of $y' = y^2$, $y(0) = 1$, $h = .05$, $n = 10$. Here $j = 3$ (the number of iterations).

```
F(X, Y) = Y * Y
H = .05
N = 10
Y = 1.
X = 0.
WRITE (6, 50) X, Y
D1Y = Y * Y
D2Y = 2 * Y * D1Y
Y1 = Y + H * D1Y + H * H/2. * D2Y
YZ = Y
X = X + H
```

```
      WRITE (6, 50) X, Y1
      DO 1 I = 1, N
      YS = Y1 + H/2. * (3 * F(X, Y1) - F(X - H, YZ)
      YJ = YS
      DO 2K = 1, 3
    2 YJ = Y1 + H/2 * (F(X, Y1) + F(X + H, YJ))
      Y = YJ
      YZ = Y1
      Y1 = Y
      X = X + H
      WRITE (6, 50) X, Y
   60 FORMAT (F10.2, E20.8)
    1 CONTINUE
      STOP
      END
```

A general method of solving $y' = f(x, y)$, $y(x_0) = y_0$ will now be given. Integrate the differential equation from x_n to x_{n+1} to obtain

$$y(x_n + h) = y(x_n) + \int_{x_n}^{x_n + h} f(x, y)dx.$$

Then use the numerical integration rule given in Chapter 6 (Equation 5).

$$\int_{x_n}^{x_n + h} f(t)dt = h(1 + \frac{1}{2}\nabla + \frac{5}{12}\nabla^2 + \frac{3}{3}\nabla^3 + \frac{251}{720}\nabla^4 + \cdots) f(x_n).$$

This yields

$$y_{n+1} = y_n + h(1 + \frac{\nabla}{2} + \frac{5}{\cdot 12}\nabla^2 + \frac{3}{8}\nabla^3 + \frac{251}{720}\nabla^4 + \cdots) f(x_n, y_n)$$

Ignoring fourth and higher differences and writing out the differences in ordinate form we have

$$y_{n+1} = y_n + \frac{h}{24}[55f(x_n, y_n) - 59f(x_{n-1}, y_{n-1}) + 37f(x_{n-2}, y_{n-2}) - 9f(x_{n-3}, y_{n-3})]$$

This can be considered as a predictor for the solution of the differential equation. In a similar manner a corrector can be derived. It is

$$y_{n+1} = y_n + \frac{h}{24}[9f(x_{n+1}, y_{n+1}) + 19f(x_n, y_n) - 5f(x_{n-1}, y_{n-1}) + f(x_{n-2}, y_{n-2})]$$

The predictor is called the Adams-Bashforth method and the corrector is called the Adams-Moulton method. We note that again starting values must be computed. Since this method has local truncation error $O(h^5)$, we would expect to use either Taylor's series or the Runge-Kutta method.

Example

Write a Fortran program to compute the solution of $y' = y^2$, $y(0) = 1$, $n = 25$, $h = .01$, for the Adams method.

```
F(X, Y) = Y * Y
X = 0.
Y = 1.
WRITE (6, 60) X, Y
DY1 = Y * Y
DY2 = 2. * Y * D1Y
D3Y = 2. * (Y * D2Y + D1Y * D1Y)
D4Y = 2. * Y * D3Y + 6. * D1Y * D2Y
H = .01
H2 = 2. * H
H3 = 3. * H
X1 = X + H
X2 = X + H2
X3 = X + H3
Y1 = Y + H * D1Y + H * H/2. * D2Y + H ** 3/6. * D3Y
                    + H ** 4/24. * D4Y
Y2 = Y + H2 * D1Y + H2 * H2/2.* D2Y + H2 ** 3/6.* D3Y
                    + H2 ** 4/24. * D4Y
Y3 = Y + H3 * D1Y + H3 * H3/2.* D2Y + H3 ** 3/6. *D3Y
                    + H3 ** 4/24. * D4Y
YZ = Y
```

```
      WRITE (6, 60) X1, Y1
      WRITE (6, 60) X2, Y2
      WRITE (6, 60) X3, Y3
      X = X + 3. * H
      DO 1 I = 1, 25
      YJ = Y3 + H/24. * (- 9. * F(X - 3. * H, Y2)
               + 37. * F(X - 2. * H, Y1) - 59. * F(X - H, Y2)
               + 55. * F(X, Y3)
      DO 2 J = 1, 2
    2 YJ = Y3 + H/24. * (F(X - 2. * H, Y1) - 5. * F(X - H, Y2)
               + 19. * F(X, Y3) + 9. * F(X + H, YJ))
      X = X + H
      WRITE (6, 60) X, YJ
   60 FORMAT (F10.2, E20.8)
      YZ = Y1
      Y1 = Y2
      Y2 = Y3
    1 CONTINUE
      STOP
      END
```

Adams method may be extended to solutions of n simultaneous ODE's but we shall not do it here.

Exercises

1. Write a Fortran program to solve $y' = y$, $y(0) = 1$, using the Adams method with $h = .05$, $n = 20$. Compare your result with the true answer $y = e^x$.

2. Write a Fortran program to compute $y' = y^2 + x^2$, $y(0) = 1$, $n = 50$, $h = .02$, using the Adams method.

Chapter 8

THE NUMERICAL SOLUTION OF INTEGRAL EQUATIONS

In this chapter we consider the numerical solution of integral equations of the form

$$y(x) = f(x) + \int_0^x k(x, s, y(s))ds.$$

This is called a Volterra integral equation of the second kind. Here f(x) and k(x, s, y(s)) are given functions and y(x) is the unknown function.

Example

$$y(t) = \cos(t) + \int_0^t \sin(t - x)\ y(x)dx$$

Here $f(t) = \cos(t)$, $k(t, x, y(x)) = \sin(t - x) \cdot y(x)$. The solution is $y(t) = 1$ as is easily verified.

$$y(t) = \cos t + \int_0^t \sin(t - x) \cdot 1\ dx$$

$$y(t) = \cos t + \cos(t - x)\Big|_0^t = \cos t - \cos t + 1$$

Hence $y(t) = 1$.

In the numerical solution of integral equations we wish a solution at a discrete set of points 0, h, 2h, 3h,$\cdots$, nh where $h > 0$. We designate the solutions by $y_0, y_1, y_2, \cdots, y_n$. We begin the numerical method by first approximating the integral on the right-hand side at $x = h$ by the rule

$$\int_{x_0}^{x_0 + h} f(t)dt \cong h\, f(x_0)$$

to obtain

$$y_1 = f_1 + h\, k(h, 0, y_0).$$

Here

$$y_0 = f_0 + \int_0^0 k(0, s, y)dx = f_0$$

$$y_2 = f_2 + h[k(2h, 0, y_0) + k(2h, h, y_1)], \text{ etc.}$$
$$y_n = f_n + h[k(nh, 0, y_0) + k(nh, h, y_1) + \cdots k(nh, (n - 1)h, y_{n-1}].$$

Example

Compute the solution of $y(t) = \cos t + \int_0^t \sin(t - x)\, y(x)dx$

with $h = .1$, for y_0, y_1, y_2.

$$y_0 = y(0) = \cos(0) + \int_0^0 \sin(- x)\, y(x)dx = 1$$

$$y_1 = \cos h + \int_0^h \sin(h - x)\, y(x)dx$$
$$= \cos h + h \sin(h)\, y = .99500 + .1(.09983)$$
$$= 1.004983$$

$$y_2 = \cos (2h) + \int_0^{2h} \sin(2h - x)\, y(x)dx$$
$$= \cos (.2) + h[\sin (2h)y_0 + (\sin h)\, y_1]$$
$$= .98007 + .1[.19867 + .09983] = 1.00992$$

Example

Write a Fortran program to compute the solution of

$$y(t) = e^{-t^2} (1 + t^2) + t + 1 + \int_0^t t^2 e^{-ts} y(s)ds.$$

Here h = .01, n = 100

```
      DIMENSION Y(101)
      REAL K
      F(T) = EXP(- T * T)*(1. + T * T) + T + 1
      K(T, S, Y) = T * T * EXP (- T * S) * Y
      T = 0.
      H = .01
      N = 100
      YZ = F(0.)
      WRITE (6, 60) T, YZ
      DO 1 I = 1, N
      S = I
      T = S * H
      Y(I) = F(T) + H * K(T, 0., YZ)
      NN = I - 1
      IF (NN .EQ. 0) GO TO 3
      DO 2 J = 1, NN
      Z = J
      ZZ = Z * H
      Y(I) = Y(I) + H * K (T, ZZ, Y(J))
    2 CONTINUE
    3 WRITE (6, 60) T, Y(I)
   60 FORMAT (F10.2, E20.8)
    1 CONTINUE
      STOP
      END
```

A more accurate scheme is to use the trapezoidal rule on the integral under consideration. Consider

$$y(x) = f(x) + \int_0^x k(x, s, y(s))ds.$$

Then for x = h we have, applying the trapezoidal rule to the resulting integral,

$$y_1 = f_1 + \frac{h}{2}[k(h, 0, y_0) + k(h, h, y_1)]$$

We must then iterate for the value of y_1. A good first guess is given by $y_1 = f_1 + h[k(h, 0, y_0)]$, the Euler method.

In general we must solve

$$y_n = f_n + h\left[\frac{k(nh, 0, y_0)}{2} + k(nh, h, y) + \cdots + k(nh, (n-1)h, y_{n-1}) + \frac{k(nh, nh, y_n)}{2}\right]$$

A good first guess for y_n is

$$y_n = f_n + h[k(nh, 0, y_0) + \cdots + k(nh, (n-1)h, y_{n-1})]$$

In case the equation is linear, i.e.,

$$k(x, s. y(s)) = k(x, s)y(s)$$

we can solve for y_n and avoid the iterations.

Example

$$y(t) = \cos t + \int_0^t \sin(t - x)y(x)dx$$

Again, $y(0) = 1$.

$$y_1 = \cos h + \frac{h}{2}[\sin h\ y_0 + 0(-y_1)]$$

Since here $k(h,.h, y_1) = 0$, we need not solve for y_1

$$y_1 = .99500 + \frac{.1}{2}[.09983] = .999991$$

$$y_2 = \cos(2h) + .1\left[\frac{\sin(2h)y_0}{2} + \sin h\ y_1 + \frac{\sin(0)y_2}{2}\right]$$

$$y_2 = .98007 + .1\left[\frac{.19867}{2} + .09983\ y_1 + 0\right]$$

$$= .98007 + .019916 = .999923$$

Thus the trapezoidal rule appears to be giving us better results for this example than Euler's method.

A Fortran program for the example

$$y(x) = 1 + \int_0^x 5y(s)ds$$

is given below. Here $h = .01$, $n = 100$, $y(0) = 1$.

```
      DIMENSION Y(100)
      REAL K
      F(T) = 1.
      K(T, X, Y) = 5. * Y
      X = 0.
      H = .01
      N = 100
      YZ = F(0.)
      WRITE (6, 60) X, YZ
      DO 1 I = 1, N
      T = I
      X = T * H
      Y(I) = F(X) + H * K(X, 0., YZ)
      NS = I - 1
      IF (NS .EQ. 0) GO TO 15
      DO 2 J = 1, NS
      Z = J
      S = Z * H
      Y(I) = Y(I) + H * K(X, S, Y(J))
    2 CONTINUE
   15 YIP = F(X) + H/2. * K(X, O., YZ)
      IF (NS .EQ. 0) GO TO 35
      DO 30 J = 1, NS
      Z = J
      YIP = YIP + H * K(X, Z * H, Y(J))
   30 CONTINUE
   35 DO 40 L = 1, 3
   40 Y(I) = YIP + H/2. * K(X, X, Y(I))
      WRITE (6, 60) X, Y(I)
   60 FORMAT (F10.2, E20.8)
    1 CONTINUE
      STOP
      END
```

Exercises

1. Compute the solution of

$$y(x) = 1 + \int_0^x (-3)\, y(s)ds$$

for the values .1, .2, .3, using h = .1 by Euler's method. Compare with the true solution $y(x) = e^{-3x}$.

2. Write a Fortran program to compute the approximate solution of

$$y(t) = 1 - t + \int_0^t \left(te^{x(t-2x)} + e^{-2x^2}\right) y(x)dx$$

using the trapezoidal rule with h = .05, n = 20, $y(x) = e^{x^2}$.

Chapter 9

LEAST SQUARES SOLUTIONS

9.1 Introduction. In the Introduction and in Chapter 4 we studied various ways of approximating a function, namely by Taylor's series and interpolation. Another way of approximating a function is to use the method of least squares. The technique is popular where experimental data are given.

Section 9.2 Continuous Data. Suppose we wish to fit the function $f(x) = e^x$ by a first degree function over the interval [0, 1] by the method of least squares. We have

$$\int_0^1 [A + Bx - e^x]^2\,dx = \text{min.} \quad \text{for all choices of A and B.}$$

We write

$$I(A, B) = \int_0^1 [A + Bx - e^x]^2\,dx = \text{min.}$$

Hence

$$\frac{\partial I}{\partial A} = \frac{\partial I}{\partial B} = 0.$$

We obtain the equations

$$\frac{\partial I}{\partial A} = 0 = 2\int_0^1 [A + Bx - e^x]\,dx$$

$$\frac{\partial I}{\partial B} = 0 = 2\int_0^1 [A + Bx - e^x]\,x\,dx$$

or the two equations

$$A \int_0^1 dx + B \int_0^1 x\,dx = \int_0^1 e^x\,dx$$

$$A \int_0^1 x\,dx + B \int_0^1 x^2\,dx = \int_0^1 x\,e^x\,dx$$

Integrating the above coefficients we have

$$\int_0^1 dx = 1, \int_0^1 x\,dx = \frac{1}{2}, \int_0^1 x^2\,dx = \frac{1}{3}$$

$$\int_0^1 e^x\,dx = e^x \Big|_0^1 = e - 1$$

$$\int_0^1 x\,e^x\,dx = x\,e^x \Big|_0^1 - \int_0^1 e^x\,dx = e - e^x \Big|_0^1 = 1$$

Thus we have the system of equations

$$A + \frac{B}{2} = e - 1$$

$$\frac{A}{2} + \frac{B}{3} = 1$$

We solve to obtain

$$A = 4e - 10, \quad B = 18 - 6e$$

Suppose we wished to fit the function f(d) to a linear least squares curve fit, then we would have over [a, b]

$$I(A, B) = \int_a^b [A + Bx - f(x)]^2\,dx$$

If we wish I(A, B) to be a minimum, we require that

$$\frac{\partial I}{\partial A} = \frac{\partial I}{\partial B} = 0$$

Thus we obtain the system of equations

$$A \int_a^b 1\,dx + B \int_a^b x\,dx = \int_a^b f(x)dx$$

$$A \int_a^b x\,dx + B \int_a^b x^2\,dx = \int_a^b xf(x)dx$$

to be solved for A and B.

Suppose we wish to solve the least squares fit for a quadratic function. Then we have

$$I(A, B, C) = \int_a^b [A + Bx + Cx^2 - f(x)]^2\,dx = \text{min.}$$

For a minimum we require

$$\frac{\partial I}{\partial A} = \frac{\partial I}{\partial B} = \frac{\partial I}{\partial C} = 0.$$

The following system of equations is obtained.

$$A\int_a^b dx + B\int_a^b x\,dx + C\int_a^b x^2\,dx = \int_a^b f(x)dx$$

$$A\int_a^b x\,dx + B\int_a^b x^2\,dx + C\int_a^b x^3\,dx = \int_a^b x\,f(x)dx$$

$$A\int_a^b x^2\,dx + B\int_a^b x^3\,dx + C\int_a^b x^4\,dx = \int_a^b x^2\,f(x)dx$$

Exercises

1. Fit $f(x) = e^x$ on [0, 2] to a first degree linear least squares fit.
2. Fit $f(x) = x^2 + 2$ on [0, 1] by first degree linear least squares fit.
3. Fit $f(x) = \sin x$ on $[0, \pi]$ by first degree linear least squares fit.
4. Fit $f(x) = e^x$ on [0, 1] by a quadratic least squares fit.
5. Fit $f(x) = \sin x$ on $[0, \pi]$ by a quadratic least squares fit.

Section 9.3 Discrete Data. Suppose we are given discrete data at points (x_i, y_i) $i = 1, 2, \cdots, n$. We wish to fit a first degree curve to the data in such a way as to make

$$I(A, B) = \sum_{i=1}^{n} (A + Bx_i - y_i)^2 = \text{minimum}$$

We require

$$\frac{\partial I}{\partial A} = \frac{\partial I}{\partial B} = 0.$$

The following system of equations is obtained

$$A \sum_{i=1}^{n} 1 + B \sum_{i=1}^{n} x_i = \sum_{i=1}^{n} y_i$$

$$A \sum_{i=1}^{n} x_i + B \sum_{i=1}^{n} x_i^2 = \sum_{i=1}^{n} x_i y_i.$$

Since $\sum_{i=1}^{n} 1 = n$ the first equation above reduces to

$$An + B \sum_{i=1}^{n} x_i = \sum_{i=1}^{n} y_i$$

Example

Suppose we are given the following table of data :

x_i	y_i
1	2
2	3
3	4
4	5
5	6

Then

$$\sum_{i=1}^{n} x_i = 1 + 2 + 3 + 4 + 5 = 15$$

$$\sum_{i=1}^{n} x_i^2 = 1 + 4 + 9 + 16 + 25 = 55$$

$$\sum_{i=1}^{n} y_i = 2 + 3 + 4 + 5 + 6 = 20$$

$$\sum_{i=1}^{n} x_i y_i = 2 + 6 + 12 + 20 + 30 = 70$$

So we have the system of equations

$$5A + 15B = 20$$
$$15A + 55B = 70$$

Solving we find

$$A = 1$$
$$B = 1$$

Exercises

1. Fit the following data to least squares linear fit.

x	y
0	54
1	154
2	227
3	359
4	401

2. Fit the following data to least squares linear fit.

x	y
1	3
2	4
3	7
4	15
5	25

It may be advisable to use a quadratic fit or some higher degree polynomial approximation to the data. We carry out the case of quadratic fit in detail here. For higher degree fits see Hildebrand [ ].

We wish to make $I = \sum_{i=1}^{n} [A + Bx_i + Cx_i^2 - y_i]^2$ a minimum. Taking partial derivatives with respect to A, B, C, we have

$$\frac{\partial I}{\partial A} = \frac{\partial I}{\partial B} = \frac{\partial I}{\partial C} = 0.$$

We obtain the following system of equations.

$$A \sum_{i=1}^{n} 1 + B \sum_{i=1}^{n} x_i + C \sum_{i=1}^{n} x_i^2 = \sum_{i=1}^{n} y_i$$

$$A \sum_{i=1}^{n} x_i + B \sum_{i=1}^{n} x_i^2 + C \Sigma x_i^3 = \sum_{i=1}^{n} x_i y_i$$

$$A \sum_{i=1}^{n} x_i^2 + B \sum_{i=1}^{n} x_i^3 + C \Sigma x_i^4 = \sum_{i-1}^{n} x_i^2 y_i$$

Exercise

1. Fit the following data to least squares quadratic fit.

x	y
.5	.47943
.7	.64422
.9	.78333
1.1	.89121
1.3	.96356